Family Farms: Survival and Prospect

Marx, Lenin and Kautsky all regarded family farming as doomed to be split into capitalist farms and proletarian labour. Most modern economists regard family farming as an archaic form of production organization, destined to give way to agribusiness. *Family Farms* refutes these notions and analyses the manner in which family farmers have been able to operate with success in both developed and developing countries, using examples wherever these are illuminating.

This book begins by reviewing theoretical arguments about agricultural structures, and defines family farming. This is followed by five vignettes about farming in the first half of the twentieth century. The authors analyse the conditions of access to land and water, labour, livestock, tools and seed and review marketing arrangements and how they have changed since 1900. A three-chapter review of evolving policies in the North Atlantic countries, in the communist states, and in the developing countries, leads to a discussion of the impact of neo-liberalism. New issues of the farmer as steward of the environment are then explored, as well as modern ideas about de-agrarianization and a discussion of land reform, tracing the experience of Mexico and Brazil. In two final chapters the more positive approach of pluriactivity is discussed and followed by a review of organic farming as a principal modern innovation. New political organizations representing family farming are described and their demands are discussed with empathy, but in a sceptical manner.

Family farming is an adaptable and efficient form of production organization, and these qualities have allowed it to survive. The future will be no easier than the past, yet family farming continues to flourish in most contexts. This book will be useful for researchers, students and lecturers interested in Development Studies, Rural Studies and Geography and Anthropology, as well as general readers who have an interest in farming.

Harold Brookfield is a geographer. His research is on rural societies in the developing countries. He coordinated the UN project on People, Land Management and Environmental Change.

Helen Parsons, an agricultural scientist trained in Adelaide, is a member of the Department of Anthropology at The Australian National University. She worked on the UN project on People, Land Management and Environmental Change.

Routledge Studies in Human Geography

This series provides a forum for innovative, vibrant, and critical debate within Human Geography. Titles will reflect the wealth of research which is taking place in this diverse and ever-expanding field.

Contributions will be drawn from the main subdisciplines and from innovative areas of work which have no particular subdisciplinary allegiances.

Family Farms: Survival and Prospect

A world-wide analysis

Harold Brookfield and Helen Parsons

LONDON AND NEW YORK

First published 2007
by Routledge
2 Park Square, Milton Park, Abingdon, Oxfordshire OX14 4RN

Simultaneously published in the USA and Canada
by Routledge
711 Third Avenue, New York, NY 10017

First issued in paperback 2014

Routledge is an imprint of the Taylor & Francis Group, an informa business

Typeset in Times New Roman by
Keystroke, 28 High Street, Tettenhall, Wolverhampton

British Library Cataloguing in Publication Data
A catalogue record for this book is available from the British Library

Library of Congress Cataloging in Publication Data
Brookfield, H. C.
   Family farms : survival and prospect : a world-wide analysis /
   by Harold Brookfield and Helen Parsons.
      p. cm.
   Includes bibliographical references and index.
   ISBN 978–0–415–41441–8 (hardcover)
   1. Family farms—History. 2. Agriculture and politics.
   I. Parsons, Helen, 1955– II. Title.
   HD1476.A3B76 2007
   338.1—dc22
   20070212219

ISBN13: 978–0–415–41441–8 (hbk)
ISBN13: 978–0–415–75960–1 (pbk)

For our partners and sons
Muriel, George, John, Max and Eddy

Contents

Boxes and Tables

Boxes

Tables

Preface

'You are writing a book about small family farms world-wide? Historical, is it? Surely the family farm is almost extinct?' The speaker was a city-dweller in a developed country, albeit in a small city surrounded by family farms. He was also an academic historian, well aware that Marx and Lenin predicted that capitalism would dissolve the peasantry into capitalist farmers on the one hand and proletarian workers on the other. Most surely, he had read Eric Hobsbawm's (1994: 289–93) summary account, backed up by employment data from many lands, of the 'death of the peasantry' between the 1940s and 1980s. He was less likely to have read Franklin (1969), who predicted that the European peasantry would have ceased to exist by about the end of the twentieth century. We have read several more recent obituaries for the small family farmer. They are written in advance of their actual demise, but there is a widespread belief that globalization has finally destroyed the basis for their independent existence so that, to cite one example, 'North American family farmers now seem to be on the cusp of virtual elimination, both as direct producers and as social actors' (Adams 2003: 11). Their replacement is seen to be industrially organized 'factories in the fields' enjoying all the economies of scale. Many believe that something very like this has already taken place in the United States, that only a massive structure of subsidies has slowed the transformation in Europe, and that it is rapidly occurring in other developed countries.

In part, this book springs from reaction to all this gloom. Small- to medium-scale family farmers have not disappeared nor do they show any immediate sign of doing so. The 1980s and 1990s, the decades in which Hobsbawm consigned them to the dustbin of history, in fact saw the emergence of small farmers' political movements on an unprecedented scale. The agro-industrial corporation, of which the real prototype was probably the tropical crop plantation, staffed by slaves or indentured labourers, has become important in certain fields of activity, but has not become the dominant mode of rural production. There are certainly many company farms, but a lot of these are family companies, incorporated for taxation convenience. In Europe, Canada, Australia and New Zealand, and even in much of the USA, the great majority of farms continue to be operated by families. They have not survived unchanged, and for most farmers in most lands, the twentieth century was a traumatic era. Although we do not set out to write a rural history of the twentieth century, the changing conditions of farming through the past hundred years are the subject of most of our chapters. We cannot fully understand the present, nor try to predict the future, without knowing how the present came about.

The other reason we had for writing this book is what we have learned from lifetime experience among farmers in many lands. This is the commonality of the conditions of family farming over a wide range of social and political conditions in the modern world. Whether we are among farmers in New Guinea, Latin America, China, Africa, North America or Europe, we are among people practising the same basic trade of producing a livelihood from the earth and its biota. All have to deal with people who have power over them as persons, over their land, over the disposal of their produce, or over all three of these. The form changes from place to place and time to time, but the basic structure does not. To say this is not to ignore the great contrasts in the world between conditions in the wealthy countries and those in the developing countries. It does put the atomized population of farmers into their widely replicated relationship of subordinate dealing with the more powerful who can control their access to markets, to the means of production and, to a greater or lesser degree, their freedom of individual decision making. Despite the radical differences in the way in which these constraints are imposed, this unequal relationship has, in the twentieth century, applied under pre-capitalist, capitalist and communist forms of governance. We show this in our chapters that follow.

This book discusses the changing conditions of family-scale farming comparatively across these different politico-economic systems, principally from the standpoint of how the conditions impact on farmers' livelihoods and, in turn, of the adaptations that farmers have been able to make. We write of family farmers in both developed and developing countries, and also of those in Russia and China. One group that would certainly agree with our view that all family farmers share a common set of problems is the membership of the international small-farmers' movement, *Via Campesina*, discussed in Chapter 14. Clearly, our approach is unusual. Few writers have attempted to compare actual farming conditions in such a varied set of lands. Bayliss-Smith (1982) is one of our few precursors down this road. His small book uses the tools of energy input/output analysis, very popular in the 1970s and 1980s, to compare the efficiency of both production and distribution systems in seven case-study communities ranging from pre-industrial to fully industrial in their production modes. Our approach, like his, seeks to avoid the perils of overgeneralization by the use of case-study material, our own and that of others. This means that we are inevitably selective. But we lack Bayliss-Smith's means for quantified comparison, and our discussion is qualitative. Our purpose is not to seek efficiencies but to understand how a very large proportion of the world's farmers obtain and allocate their resources, dispose of their products, and relate to other sectors of society and economy around them.

These related sectors are of major importance to our analysis. While it is statistically right to point to the declining role of agriculture in the national economies of all developed countries, this is to isolate the farm from those who supply the means of production, and those who buy and often manufacture from the farm product. While farming itself accounts for only around 2 per cent of gross domestic product (GDP) in these countries and much is made of this in general economic literature, the true comparison with other sectors should include all associated activities, even down to the worker on the supermarket checkout who performs much the same role in the system as the farmer's wife or daughter selling produce by the roadside or in a village market in an African country. The farm and food sector as a whole makes up from 10 to 20 per cent of the whole economy in a range of countries, including Britain, the USA and Australia. Farmers and their staffs, as the production workers in this system, have a central place in very large economic sectors.

There is a number of ways in which our subject could be approached. We are grateful to anonymous referees for the publishers who dissuaded us from persisting with one that was clearly inappropriate. Although the material is presented as a continuous flow the better to facilitate backward and forward reference, there are fairly distinct sections in the presentation. After two initial chapters which face the 'agrarian question', define the family farm and provide a backward glance at farming in the early twentieth century, chapters 3, 4 and 5 set out the basics of working the farm and marketing its produce. Chapters 6, 7 and 8 then introduce the relationship of farmers with the state, successively in what we call the 'North Atlantic' countries, then Russia and China and third, more generally, in the developing countries. In Chapter 8 we specifically introduce neo-liberalism. Chapters 9 and 10 next review some modern trends and the new role for farmers as environmental custodians for their fellow-citizens. In Chapter 11 we look at the question of land reform and in Chapter 12 the large question of de-agrarianization. Chapter 13 examines the more positive approach of pluriactivity, and we discuss the recent growth of organic farming. Finally, in Chapter 14, we examine prospects through a generation span in the twenty-first century. Our conclusion is mixed, but not pessimistic.

We make very limited use of the term 'peasant' in this book. This is because of the unfortunate pejorative implications of the term 'peasant' in colloquial English, absent from other languages. Lehmann (1996) reviewed the several meanings of 'peasant' in English and advised that the term be dropped. All our farmers would in French be *paysans*, in German *bauern* and in Spanish *campesinos*, but in English they have to be family farmers, or just farmers. In Chapters 8 and 11, where we are writing of Latin American farmers, we use campesinos.

Although we have written this book in a department of anthropology which has been very kind to us, we are by training a geographer and an agricultural scientist. One of us was both brought up on a farm and lives on one now, while the other has only some teenage experience of actual farm work. Both of us, however, have had extensive experience of work with and among farmers in the developing countries and this, probably, is the experience that comes through most strongly in our writing. From the early 1990s until its end in 2002, we were intimately involved in the work of a UN project on farmers as guardians of biodiversity, entitled People, Land Management and Environmental Change (PLEC), referred to at several points and described in Chapter 10. Most of our work has been concerned with what farmers do, with their management practices in relation to the environment. Most recently, we have written, edited or worked on three books concerned in one way or another with the diversity of what they do and of the landscapes that they produce (Brookfield 2001; Brookfield *et al.* 2002; Brookfield *et al.* 2003). In none of these books, nor in the UN project, did we deal adequately with the social and political contexts in which their work is embedded. This book addresses that gap.

Acknowledgements

Our principal acknowledgement is to those who worked with us in the UN University project on People, Land Management and Environmental Change (PLEC) between 1992 and 2002, who introduced us to many of the issues and places discussed in this book. We are grateful to the Department of Anthropology, Research School of Pacific and Asian Studies, The Australian National University, in which the book was written and which has provided us with constant material support. Within it, we thank particularly Fay Castles, the department's administrator, and Chris Thomson, its IT specialist who has patiently helped us with many problems. In Routledge, Zoe Kruze and Jennifer Page have offered constant support, and we are most grateful to two anonymous referees who commented valuably on our proposals, and led us to restructure the plan of writing. Muriel Brookfield, who worked with us in the PLEC project, consented to read through and comment on the chapters once we had brought them to a near-final state, and has saved the reader many infelicities. No one else but the publisher's experts has read the whole manuscript. We got some good ideas from our colleagues in response to a seminar in 2005, and Andy Kipnis made valuable comments on Chapter 7. John Connell told us about central Madagascar in 2005. Only we are responsible for the errors.

Canberra, Australia and Quillabamba, Peru, July 2007

Abbreviations

AES	agri-environmental scheme
ASERCA	*Apoyo y Servicios a la Comercialización Agropecuaria*
AUD	Australian dollar
BSE	bovine spongiform encephalopathy
CAFOs	concentrated animal feeding operations
CAP (EU CAP)	Common Agricultural Policy of the European Union
CBD	Convention on Biological Diversity
CCC	Commodity Credit Corporation
CIDA	*Comité Interamericans de Desarollo Agrícola*
CONASUPO	*Comisión Nacional de Subsistencias Populares* (Mexico)
CPE	*Coordination Paysanne Européenne*
CRP	Conservation Reserve Program
DDT	Dichloro-Diphenyl-Trichloroethane
EAFRD	European Agricultural Fund for Rural Development
ECLA	Economic Commission for Latin America
EQIP	Environmental Quality Incentives Program
EU	European Union
FAO	Food and Agriculture Organization
FELDA	Federal Land Development Authority (Malaysia)
FMD	Foot-and-mouth disease
GATT	General Agreement on Tariffs and Trade
GDP	gross domestic product
GEF	Global Environment Facility
GM	genetically modified
GNP	gross national product
GTV	*Gestion des Terroirs Villageois*
IFAD	International Fund for Agricultural Development
ILO	International Labour Organization
IPM	integrated pest management
IUCN	International Union for the Conservation of Nature and Natural Resources (The World Conservation Union)
LFA	less favoured areas (in the European Union)

MAB	Man and the Biosphere Programme
MST	*Movimento dos Trabalhadores Rurais Sem Terra*
NAFTA	North American Free Trade Agreement
NFFC	National Family Farm Coalition (United States)
NGO	non-governmental organization
NPK	Nitrogen:Phosphorus:Potassium
OECD	Organization for Economic Cooperation and Development
PAN	National Action Party (Mexico)
PLEC	People, Land Management and Environmental Change
PRI	Party of the Institutionalized Revolution (Mexico)
PROCAMPO	*Programa de Apoyos Directos al Campo*
PROCEDE	*Programa de Certificación de Derechos Ejidales y Titulación de Solares Urbanos*
TVEs	township and village enterprises
UK	United Kingdom
UNEP	United Nations Environment Programme
UNESCO	United Nations Educational, Scientific, and Cultural Organization
UNU	United Nations University
USA (US)	United States of America
USDA	US Department of Agriculture
USSR	Union of Soviet Socialist Republics
WHO	World Health Organization
WTO	World Trade Organization

1 Asking agrarian questions: defining the family farm

The total disappearance of the family farm has been confidently predicted for almost a century and a half, and is still predicted today. While a great number have not survived into the twenty-first century, the fact that so many have done so, and in so many different lands, is remarkable. In what is now termed 'late capitalism' the continued existence of these non-corporate units of production seems anomalous, even archaic, to many observers. An integrated explanation is offered at the end of the book, but in this first chapter we show why they were expected to become extinct in a much earlier phase of capitalist evolution. Then we define what we mean by 'family farm'. Some other issues of importance are introduced, to be more fully developed in later chapters.

Capitalism and farms: the classic 'agrarian question'

The model

In Britain, between the fifteenth and eighteenth centuries, the characteristic farmer became no longer a peasant working strips of land received in return for services or quit-rent to a lordly holder, but a tenant, paying rent on leasehold land and working it himself or with the aid of hired labourers. Meantime, land had become the property of a growing class of persons who included descendants of feudal lords, but were also operators of a range of businesses. Labour was no longer tied to the provision of services to the lords but now dependent on wages and became 'free' to move into other fields of activity. This process took a long time, but it advanced faster and further in Britain than in other countries. The dissolution of the old feudal bonds, and the creation of a 'free' population of workers, also created a population of capitalists who accumulated that part of the product of labour not necessary to be paid to the workers for their survival needs, called the 'surplus value' of labour. They were themselves free to invest the accumulated surplus in new industries, and to compete with one another in the market. Adam Smith (1776: 155) had already noted that the 'surplus produce of the country . . . over and above the maintenance of the cultivators . . . constitutes the subsistence of the town'. This necessary relationship, reinterpreted by Marx (1867), provided the basis for explaining and (imperfectly) predicting the future of the world-wide capitalist economy.

From Marx, via Lenin and Kautsky, to the scale of farms

To Marx, who was simultaneously economist, sociologist and historian, the initial basis of capitalism was 'primitive accumulation' of capital. Following the English model, as he not altogether accurately interpreted it in the famous Chapter 24 of *Capital I* (Marx 1867), this was done through enclosures, in the context of a growing market economy, involving the separation of the worker from control of the means of production, whether as owner or renter. This, in turn, required production of commodities traded on the market, using paid labour. On the land, therefore, capitalists acquired the means of production, and labour became proletarian.[1] The competitive discipline of 'market dependence' led to the growth of productivity of both labour and land, by means of improvements and innovations. This same growth in productivity lowered the costs of providing food and other raw materials for the growing urban and other non-agricultural population created by capitalist transformation. The 'agrarian question' of the nineteenth century concerned how surpluses earned from agriculture are transferred to investment in other economic sectors, specifically industry (Kautsky 1899; Lenin 1899). Capital acquired from farmers by merchants and usurers was one major channel. Another was 'unequal exchange', being the difference (in labour value) between returns for agricultural and industrial goods. A similar basic question came to dominate much non-Marxist economic thinking about developing countries in the twentieth century. The most common solution was to enlarge primary produce exports, taxed to provide the foreign exchange needed to fund industrial development.

While accumulation provided the resources needed for industry, and the labour 'freed' from the land also provided the labour for industry, there were important consequences among the agricultural population. These consequences happened whether formerly feudal lords converted themselves into capitalist large farmers, or whether such farmers arose by differentiation among the rural landowners. All aspects of the capitalist economy are competitive.[2] Following the internal differentiation approach, and elaborating greatly on Marx, Lenin (1899) argued that competitive success would raise the wealthier peasants to the status of capitalist farmers while the less successful would fall to the level of proletarian labour. In the process, the 'middle peasant' would disappear over time. He illustrated these arguments from a large body of original data on Russian farming. The principal data were from inquiries carried out for local and provincial councils (*zemstvo*) set up in 1864 after the emancipation of the serfs in 1861. Lenin made principal use of data from ten of these inquiries. His research in the *zemstvo* data demonstrated that substantial differentiation had arisen among Russian peasant farmers, in land held and worked, livestock owned and farm machinery employed. He found what seemed clear evidence of 'capitalist contradictions' within the Russian village economy.

Kautsky (1899) cast his net more widely over Europe and found that the liberal economic revolution of the nineteenth century had not widely led to the breakdown in peasant societies that Marx had anticipated. Following Marx in treating the emergence of capitalism as primarily in manufacturing, he viewed the emergence of capitalist/ proletarian division in the countryside in the context of already-established rural–urban trade. Market competition would favour the larger-scale producer who was better placed to adopt the many technical innovations employing wage labour. The smaller peasant would be squeezed out by this competition. But this would take time because farms grow larger and smaller for a variety of independent reasons. Kautsky, and many others who have used the same argument, saw the great advantage of the large farmer as lying in the economies of scale in production, in agriculture just as in industry where Marx and other earlier writers had seen them already. A capitalist farmer could afford to mechanize, deploy labour most efficiently, attract and apply credit, and make better use of scientific advances than a small farmer. Given these advantages, the large farmer could produce more efficiently. These competitive advantages, widely perceived to this day, also underlie the strong preference for capitalist farming exhibited in the modern era of neo-liberal free-market economics. But there have been long and profound debates on the question of the comparative efficiency of large and small farms. Yields per hectare are often higher on small farms than on larger farms. While a lot of the modern argument rests on a now outdated comparative study by Berry and Cline (1979), the yield advantage of smaller farmers has more recently been argued again, principally from modern data in Brazil, by Griffin *et al*. (2004). As Byres (2004) argues in response, it is necessary to distinguish between yield per hectare and yield per unit of labour, and on the basis of the latter the capitalist farm would seem to have clear advantage. Writing of a modern form of agrarian question, that we come to only toward the end of the book, Bernstein (2006) is at pains to point out that scale by itself no longer seems a major consideration in – at least scholarly – debate.

Friedmann, Chayanov and the viability of the commercial family farm

In a series of papers arising out of her doctoral research on the social history of modern wheat farming, Harriet Friedmann (1978a, 1978b, 1980) called attention to an aspect of events in the late nineteenth century and subsequently that had been surprisingly neglected. Kautsky (1899) had noted that the non-capitalist 'middle peasant' in Europe had shown considerable resilience in the last quarter of the nineteenth century, but regarded this as only temporary. It was not. After American wheat started to flow substantially into the European market in the 1870s, and prices declined worldwide, capitalist farms in both Europe and America gave way to family-operated farms, not the other way around. By 1935, after the second major period of low prices, the vast majority of commercial wheat production throughout the world was organized through household rather than wage labour. Yet by the 1870s capitalism had reached all wheat producing regions involved in trade, and in most of them there were already capitalist farms dependent on wage labour for production. Household and wage labour farms were therefore in direct competition, and in all this long period, it was the latter that yielded ground.

Friedmann's complex explanation is in part technical.[3] In the great plains of the USA, harvesting and, later, threshing machinery were rapidly being introduced, reducing the labour requirement on farms. As family farms grew in area by purchase and rental, so machinery suitable for the new average area (about 140 ha after 1920) also became available, making mechanized farming feasible for a family labour force averaging 1.5 male persons; data on female workers were, and usually remain, insufficient. While new land remained available, and with it credit, labourers were drawn off the large capitalist farms of the 1870s and 1880s to set up family farms. The large farms, requiring both to pay rising wages and to earn profits, were unable to compete with household farms that experienced neither of these constraints.

Capitalist farmers in Europe had to compete with American, Canadian and later also Australian and Argentinian producers using household labour at received wheat prices, which converged rapidly between countries and all moved together after the mid-1890s. German capitalist farmers, who for the most part had been feudal landlords in earlier days, sought to survive by coercive labour management and importation of Polish labourers; like French farmers, they were assisted by tariff protection. British farmers, who were not assisted in this way, mostly shifted out of arable production into livestock specialization, reducing their need for labour. In the British case, unlike the German, most farmers were already tenants and the decline in land values was borne by the non-farming owners of the land.

The technical conditions were important, but needed to be considered in conjunction with the contrasted social conditions of production in capitalist and household labour farms. Here Friedmann drew on the work of Aleksandr Chayanov (1923, 1925, 1966), who had used the same *zemstvo* data as Lenin, but relied also on inquiries by himself and his colleagues and students. Chayanov found the demographic stage of a household, as it first grew and then declined after the young became independent, to be dominant in bringing about differentiation between family farms, whatever their degree of commercial orientation.[4] He concluded that the balance between consumers and workers within the household was the main determinant of the scale of production. In Table 1.1 we adapt his oft-quoted (and oft-misunderstood) table summarizing results in the district of Volokolamsk, west of Moscow.

Chayanov's explanation relied on the fact that household workers consume the product of their work, whether as food and clothing or as money, but are paid no direct wages. Chayanov isolated the family farming household from its surrounding

Table 1.1 Productivity and intensity of work in relation to household composition at Volokolamsk, 1910

	Consumers per worker			
Consumer/Worker ratio	1–1.2	1.21–1.4	1.41–1.6	over 1.6
Worker's output (roubles)	131.9	151.5	218.8	283.4
Working days per worker	96.8	102.3	157.2	161.3

Note: Based on data for 25 household farms, after Chayanov (1966: 78, table 2–8).

partly capitalist economy for theoretical purposes. He treated it as 'a family that does not hire outside labour, has a certain area of land available to it, has its own means of production [i.e. tools, etc.], and is sometimes obliged to expend some of its labour force on non-agricultural crafts and trades' (Chayanov 1966: 51). There is no structural requirement for the farm to make a profit. The household is treated as a single, undifferentiated, producing and consuming unit. It will produce, or earn from work off the farm, what is needed to pay its rent (if any) and taxes, keep the farm functioning (in Marxist terms 'reproduce' it) and satisfy its own demands, but will not willingly do more than this. The consumer/worker balance of the family farm will change primarily through demographic process as families enlarge, grow old and are replaced. In effect, Chayanov was relying on the marginal analysis of neo-classical economics, so that an equilibrium level will be found where the marginal utility of outputs equals the marginal disutility of work (Hunt 1979).[5]

Collantes (2006a) usefully points out that, in comparing costs, returns and responses among farms operating under different production systems, Kautsky and Chayanov were in fact both using a neo-Darwinist evolutionary approach, formalized by Lawson (2003) as a population–variety–reproduction–selection model, in which market competition leads to selection of the 'fittest varieties'. Kautsky worked with the commodity market, and saw the wage-paying capitalist farm as the fittest; Chayanov, also considering ability to compete in markets for the factors of production, saw the peasant farm as the fittest, and Friedmann essentially agreed with him. Commenting only on Chayanov, Ellis (1988a) drew attention to revisions introduced by the 'new home economics' which arose in the 1960s and 1970s, initially through recognizing that households not only farm or refrain from farm inputs to seek leisure, but they also produce their own utilities, the use values being obtained ultimately from their final consumption. In the presence of a labour market, inputs are determined not so much by preferences but by the going wage rate and price level, which yield opportunity costs of time spent on different activities. Contrary to Chayanov's model, this allows decisions with respect to labour use to be separated from decisions with respect to income. For example, a rise in the going market wage will lead to a decline in hired labour use, a rise in farm work performed by the family and a rise in the proportion of output consumed at home. The presence of a labour market alters the internal logic of the household model and, not least, the way the household interacts with the wider economy. Ellis (1988a: 139) concludes that 'the unique mode of economic calculation proposed by Chayanov disappears'.

Yet the Chayanov family always did operate in the presence of a labour market. It was optimizing conflicting utilities in a variable environment of nature and the market. In an entertaining discussion of Marxist work on transitions to capitalism among South American peasants, illuminated by his field work in Ecuador, Lehmann (1986) writes of 'capitalist family farms' which rely on family labour and supplementary labour recruited mainly through kinship ties, but also invest in machinery specifically to avoid the need to hire labour. He is describing modern family farms everywhere in this context. Sivakumar (2001: 42) prefers to regard Indian family farmers as continuously 'adapting themselves to changing objectives and constraints', as finding not the best but a satisfactory solution, that is as 'satisficing' in modern terminology. They are managing well within production-possibility limits. On this basis, Sivakumar finds Chayanov's arguments more persuasive than either Marxian reasoning or the modern complex of ideas arising from the application of classical and neo-classical economics to agrarian issues. Sivakumar does not discuss the 'efficient-but-poor' peasants of Schultz (1964), modified under risk to be following a 'survival algorithm' (Lipton 1968). Nonetheless, he absorbs these notions within a discussion of the deep uncertainties of agriculture. Coming from a background of life and research in southern India, Sivakumar (2001) reinterpreted the decision-making problem in the context of 'transaction regimes' in which dominance and dependence, hierarchy and location in regard to the geographical distribution of resources and their variability were the fundamental context of all decisions. Both the Marxist emphasis on conflict of interest between rich and poor, and the anonymity of the competitive market economy, became special cases of transaction market construction. He concluded, therefore, that a theory of non-capitalist economic systems, such as that sought (but only partly achieved) by Chayanov, 'might better explain the agrarian situation in much of the Third World today' (Sivakumar 2001: 54). Chayanov's explanation was lost in its day under a deluge of Marxist opposition, and latterly has struggled under neo-classical criticism, although taken up by some neo-marxists to help resolve contradictions in their own arguments (Lehmann 1986). Its simplicity is its strength, and it has been taken up in modern times in a large literature, some of it well outside the range over which Chayanov was arguing, as Box 1.1 shows. The real problem concerns the type of farm, and farming environment, to which Chayanov's argument refers.

The strength of the family farm type of organization can be explored further, following Friedmann's (1980) analysis. Fully commercial household farms benefit from price improvements in just the same way as capitalist farms, and are able to invest in labour-saving machinery and other innovations which can enlarge their production. Income declines, on the other hand, are absorbed in different ways. The family farm cannot readily reduce its labour force, although individual members may temporarily leave it to work for wages, or take jobs that can be reached from home. They can reduce investment in tools, machinery and structures, but at a long-term cost if any expansion is planned. They can and do reduce personal consumption, and/or increase their inputs of work – exploiting themselves in Chayanov's argument. They can also do both these things if they want to accumulate resources to enlarge their scale of operation, or buy new machinery. This flexibility gives them a major competitive advantage over farms with a wage bill to pay, and the need for a profit. To Friedmann, the international competition in the wheat market between the 1870s and the 1930s was a competition between capitalist and household farms wherever located, in which the household farms triumphed.

The agrarian debate revised

Findings such as those of Friedmann, and the arguments of Mann and Dickinson (1978), who saw problems for capital on the land in the seasonal disjuncture of

---

**BOX 1.1 SOME UNINTENDED CONSEQUENCES OF
THEORIZING**

Kautsky, Lenin and Chayanov were not writing for modern social scientists. However, in the 1960s and 1970s, the writings of all of them became of major importance to discussion about agriculture in both developed and developing countries. Particularly remarkable has been the manner in which Chayanov's theory has been used in regard to developing country farmers. Netting (1993: 297) described the attractiveness to anthropologists of 'Chayanov's model of a primarily self-sufficient subsistence farm without wage labour and a household dedicated to its own reproduction'. It is true that Chayanov did abstract such a 'natural' farm for theoretical discussion, but in his main text he wrote that 'the subject of our analysis is precisely [. . .] a farm which has been drawn into commodity circulation' (Chayanov 1966: 125) and elsewhere (p. 119) that 'the peasant farm is acquisitive – an undertaking aiming at maximum income'. His data and most of his argument concern such farms.

Marshall Sahlins (1972) used Chayanov's finding that intensity of input on an independent family farm varies with the consumer/worker ratio to establish a foundation for his 'domestic mode of production'. Ellis (1988a) is dubious about the 'domestic mode of production', arguing that a mode of production is something that emerges from the 'social conditions of production' as a whole, and hence arises from society, not from the assembly of individual and widely differing cases. Netting (1993) was more careful than Sahlins in his use of Chayanov, but he could have read more closely. He wrote of 'land abundance' in the Russian case, where Chayanov was writing mainly of land availability, and noted large contrasts between different regions. Netting (1993) went on to read other Chayanovian characteristics as being 'self sufficiency, little commodity production or market participation, and no hired labour' (Netting 1993: 311). To make their points better, both Sahlins and Netting reproduced a simple table from Chayanov (1966: 78, table 2–8), which we also reproduce as Table 1.1. But the data in this table do not represent any sort of primitive self-sufficient economy. They derive from 1910 data on a district which showed the highest average rate of net productivity (measured in money terms) per annual worker in 'labour agricultural economic units' of any of 16 Russian districts (Chayanov 1966: 85, table 2–14). We will return to this district in Chapter 2.

---

input from output, ought to have led to a reframing of the agrarian question. There have been further detailed analyses of how family farming manages its own 'reproduction' in the most capitalist of contexts, such as that of Roberts's (1996) analysis of the southern high plains of the USA. But Marx, Lenin and Kautsky are not that easily laid to rest. The original form of the question was spelled out again in some detail by Bernstein (2004), responding with hostility to new proposals for redistributive land reform made by Griffin *et al*. (2002). Meantime, the squeezing of the farm has been viewed in the growing new dimension of agribusiness. Capitalists have found more attractive areas for investment than in agricultural production itself by providing inputs for farming and in manufacturing and dealing with outputs. They have therefore invested in agribusiness. Family (and small capitalist) farms survive as the producers who take the risks while trading with companies that are secure in more controllable fields of business. Increasingly, agribusiness controls farmers' activities (Goodman and Redclift 1985; Goodman and Watts 1994).

Chayanov, like Lenin, recognized the impact of what was then called 'merchant capital' and 'usury', both by merchants and wealthier farmers, on small-scale Russian family farmers. He proposed the vertical integration of cooperative marketing as a means of isolating the farmers from these pernicious effects, and simultaneously giving the farmers the advantages of scale in trading. In this, he drew on his own substantial experience of cooperative marketing in the difficult conditions of World War I. Processing and marketing cooperatives have proved to be an effective way of overcoming the weakness of the unsupported individual farmer in some European countries (Lamartine-Yates 1940). They have not succeeded everywhere, requiring a distinctive regime of transaction that cannot readily be produced where it does not arise spontaneously. In Russia, the horizontal integration of state and collective farming was the preferred approach, and under recent neo-liberal regimes cooperation smacks too much of a discredited socialism to be a popular path. Chayanov's analysis did not lack dynamism, but it took little account of classes. Writing only of the 'middle peasants', he treated the whole farming population as homogenous, other than the 20 per cent – Lenin's estimate – who operated in a capitalist manner. His preference for voluntary cooperatives as a way forward was contrary to the views of Stalin. It led to accusations that he was seeking to sabotage national farm production. He was arrested in late 1930 with several of his colleagues, including the distinguished Kondratieff of 'long cycle' fame, and died in 1939, still in the Gulag.

Collantes (2006a) drew on another source of ideas in discussing the competition for labour between agriculture as a whole and other economic sectors. Discussing the penetration of capitalism in the developing countries, Wolf (1982) shows how capitalist enterprises drew labour out of agriculture because the modern non-agricultural sector offered the prospect of higher living standards, easier work routines and the availability of leisure pursuits, rather than just higher incomes. Given this perception, a family farming society might be able to compete with capitalism in matters of commodity prices, capital and land markets, and yet break down due to its failure to retain labour. We see the effect of this force for change when we come to consider de-agrarianization in Chapter 12, and then in our concluding arguments. For now, the most significant modern agrarian question, quite simply, is how, and under what conditions, has the family farm survived? The next step is to discuss what might seem obvious and simple, but is not – to define the family farm.

The family farm

The stubborn 'middle-peasant' class of Lenin (1899) and Kautsky (1899), neither capitalist nor proletarian, consisted of family farmers. The family farm occupies a place between the large industrially organized 'capitalist farm' and the small allotments of paid workers employed in agriculture or elsewhere. Whether as owners or tenants, many household-scale farmers retain individual control over the land that they work. This applies both to those who are described as 'family farmers' in developed countries and to those still often called 'peasants' in developing countries, where the principal distinction is that commodities produced for subsistence are usually at least as important as commodities grown for sale (Bryceson 2002: 727). 'Peasant farms' are usefully defined by Frank Ellis (1988a: 12) as 'farm households, with access to their means of livelihood in land, utilizing mainly family labour in farm production, always located in a larger economic system, but fundamentally characterized by partial engagement in markets which tend to function with a high degree of imperfection'. Only the last of these criteria truly distinguishes family farms in the developing countries from those in the developed. Family farms also include what used to be called 'part-time farms', nowadays more often 'pluriactive' farms, on which farmers and their families rely principally on non-agricultural or off-farm sources of income. To us, all these are family farms.

Family farms are no uniform group. They may operate commercially or with only partial involvement in the market. Commercial family farms differ principally in that they are fully engaged in the market and they very often employ some labour as wage workers or contractors in addition to family members. There is a continuum from one group to the other, and in seeking to understand the future viability of the family farm we will need to take this continuum into account. We pay little attention to corporate farms, even though in the USA, Britain, the Netherlands and some other developed countries these include what are truly family-owned farms that have been incorporated mainly for taxation convenience. There are other farms in which family ownership is in some measure fictional, family members being merely prominent shareholders.[6] Farms in this group employ labour and are rarely family operated; most employ a manager. Always, however, there is a large 'shadow-zone' and no clear and obvious distinction. We recall the case cited by Blaikie and Brookfield (1987: 229–30) of the US Deputy Secretary of Agriculture in 1986, whose official position was going to debar him from receiving subsidies of about $1,000,000 to his 'family farm' in the states of California and Arizona. In his case, the term 'family farm' was surely a misnomer, but he does represent an upper bound to the scale.

Family and capitalist farms

To confine the definition of a family farm to enterprises that do not hire outside labour would determine that such farms must be small in relative terms. This is too simple, because many small farms hire from one to a few labourers without becoming anything like capitalist enterprises. It may become necessary to hire labour because of age or infirmity of family members, or because of their absence away at work. It may just make better sense to spend income in this way rather than work very long hours on the farm, or even to work at all. In Chapter 2 we describe how some Chinese male 'farmers' in 1938–39 were able to spend their whole time in leisure pursuits by taking advantage of labour available at low wages and also of the unpaid labour of their wives and daughters. Yet they were anything but capitalists.

First noting that over 98 per cent of all west-European farms are owned by individual persons, Hill (1993) made useful progress in identifying what is operationally a 'family farm'. He used the European Union database of annual work performed on farms, drawn up for 1989. Farms on which more than half of the annual work input was provided by hired workers who were not members of the farm operating family were non-family farms (in other terms, capitalist farms); only those on which 95 per cent or more of annual work input was provided by the family members qualify as 'pure' family farms. The others, with family annual work input between 50 and 94 per cent, are in an intermediate class. Together, family and intermediate classes of farms comprised 93.3 per cent of farms in the twelve 1989 European Union countries. Only in Britain and Spain did 'non-family' farms, by Hill's definition, amount to more than 15 per cent of the total.

Use of official data has problems, for countries adopt different threshold levels for recording farm data. Very small farms are generally excluded but, as Hill notes, virtually all those excluded would be in the family or intermediate class. Statistical thresholds may be changed as has happened in Australia. Data are based on an 'estimated value of farm operations' and up to the mid-1980s the threshold value was AUD 2,500. Then in two jumps it was raised to AUD 22,500 by 1996. This led to a substantial reduction in the recorded number of farms (Garnaut and Lim-Applegate 1998). Even so, 99 per cent of arable, livestock and dairy farms still recorded in 1996 were operated by an owner-manager and few made substantial use of hired labour.

Applying common sense to Hill's valuable efforts, we regard family farms to be those which are owned or tenanted, and in all cases managed by a family, and on which something like half or more of the annual work input is carried out by household members or other unpaid helpers (mostly neighbouring farm household members working on a reciprocal basis). Many family farms rely on paid labour or contract specialists at peak periods and for specific skilled tasks. Some use casual labour, or gang labour discussed in Chapter 4. Given the wide distribution of such family farm operation outside the areas of state farming, surviving collective farms and industrial plantations, we believe we are therefore writing about the large majority of all the world's farms, diverse though they are in most other respects. Moreover, a historical view would suggest a tendency for family farm operation to increase rather than decline; use of regular unskilled farm labour is generally much reduced and still in decline. Wage labour is expensive in the developed countries, and while large capitalist enterprises necessarily have to meet these costs, smaller-scale operators are decreasingly able to afford regular wage labour.

Management of family farms

While many farmers own their land, at least on mortgage or its equivalent, they rarely have full rights to determine how it is to be used. National or more local laws restrict their choices, and choice is further restricted by their agribusiness partners. Tenants have less freedom than owners to make changes unless the law protects their rights, and especially their rights to improvements. The range of variation in land access rules is enormous, and frequently it happens that the actual practice differs from the rules. Conditions of tenure and access change through time, so that 'customary rights' can become, for all practical purposes, individual and inheritable tenure. All this is discussed in Chapter 3.

Management of the individual farm is often only notionally in the hands of the senior male, in what is often described as a patriarchal pattern. Female-headed farming households are very common in areas where men work in other occupations. The farmer's immediate family is usually involved in decision making, and the farm is better thought of as a household enterprise, following Netting (1993). Households, in this sense, are defined as the people who live together, in one building or a group of buildings, and who share consumption of the outputs of the farm, in cash or kind. Sometimes more than one nuclear family is involved. In a Bambara village in Mali in the 1980s, 86 per cent of the population was in complex households containing more than one married couple, in some cases several (Toulmin 1992). In this exceptional case the largest households numbered up to 60 people. We need also to remember that the family households discussed by Chayanov were larger than those of Russian or other European farms today.

While the members of a household may characteristically be related by kinship or affinity, this is not always or necessarily so. A more difficult question concerns cases where there are associated groups of related or unrelated individuals who have patron–client relationships that can endure through time, creating continuing dependence between families who live and work on the same farm, although usually in different buildings. Japanese joint-family farms before the twentieth century normally included a landholding core, with branch farms of affinal and cognatic relatives and unrelated bond servants of different types (T.C. Smith 1959).

The dynamism of households, stressed by Chayanov, is of major importance. Natural families pass through a cycle from initial establishment, growth as children are born, and decline as children become adults and leave the family household. But individuals may leave the household to live and work elsewhere for periods of months or years without formally dissolving the household group. There is often a close relationship between the household cycle and the cycle of a farm enterprise from foundation through growth to a static or maybe declining stage. Sometimes, as on the southern high plains of the USA, the farm as a whole is not transferred or inherited, and the next generation has to reconstitute its own farm enterprises, thus renewing both family and enterprise cycles (Roberts 1996). This renewal takes place over time, normally about the span of half a generation, and not suddenly at the death or retirement of the previous holder.

How big is a family farm?

Here we reach a crucial issue. If a farm is to be operated for all regular purposes and for most of the time mainly by family labour, it cannot be of larger size than the family can manage. It was machinery that made possible substantial and real enlargement of family wheat farms in the American plains in the period between 1870 and 1935, and has continued to make enlargement possible since the mid-twentieth century. But at any technical level, there are always limits. Farm size and farm enterprise are related. Rearing of livestock can be managed with a smaller labour force over larger areas than arable production, so long as any short-term labour-intensive activities can be contracted out. Inefficiencies will always appear if the farm is too large.

It is also important that the farm be large enough, and its enterprises sufficiently demanding, to provide work for the available and willing family labour. The conditions vary greatly from place to place, as well as through time. Netting (1993) showed that there is a very regular correspondence of household size with land size wherever the household is the main body for organizing work and income rather than for consumption alone. He shows this far beyond his self-limited frame of reference to smallholders.

Normally, it is the lower bound of the family range that is transgressed, and this transgression is widely associated with poverty. In rural areas, anywhere in the world, holdings substantially less than one hectare can rarely provide the whole subsistence of even a small family. Yet they may be regarded as farms by their operators. In some Asian rice growing regions about one hectare is thought of as the minimum size for a truly viable family farm, but many do not reach this size. On the other hand, there are family farms in the Amazon delta of Brazil that occupy 30–40 ha of fields, fallow and forest, and yet are thought of as small. Ideas of optimum farm size change through time. In Europe, as late as the 1950s, it was generally thought that 10 ha was about the minimum size for viable family farm operation, when 70 per cent of farms in a group of 15 European countries were below this size (Lamartine-Yates 1960). Lamartine-Yates argued that an average of 15 ha should be the aim of policy in the second half of the twentieth century, but by 1989 an average family farm in Germany, as strictly defined by Hill, had a 'utilized agricultural area' of 26 ha. This was a little larger than the west-European average, but distinctly smaller than in Britain where the average area for such farms was 66 ha (Hill 1993). The range is discussed in other contexts in Box 1.2.

Toward a flexible and dynamic definition

In regard to scale, the term 'family farmer' necessarily has to be relative, and be opposed to the usually capitalist 'large farmers' in whatever region we are examining. In most of Latin America we would be referring to the *campesinos* as distinct from the *hacendados* or *latifundistas*. In much of sub-Saharan Africa, on the other hand, all members of many communities might be family farmers, even including the chiefs. Size of family farms is variable within a wide range. The forms

---

BOX 1.2 FARM SIZE IN OTHER LANDS

In southeastern Australia the minimum size of a family farm would be about 50–100 ha, depending on location. The upper limit is in the range of several thousand hectares. A farm in central southern Queensland is owned by two brothers. It occupies 8,000 ha, worked by them as one single farm with the use of large-scale machinery. Often, in Australia as in other countries where self-provisioning is of only minor significance, it is the size of farm income rather than the size of land that determines who is regarded by others as a 'small' farmer. Ellis (1988a) cautions against treating farm area as synonymous with economic size of farms as units of production, measurement of which would require information on the total volume of resources used in production. Like almost everyone else, he settles on farm area as the only useful comparative measure.

Change in farm size can be quite rapid, and it is not always easy to interpret. In the territory of one small subclan in the Chimbu area of highland Papua New Guinea, Brown *et al.* (1990) found that the recorded and mapped cultivated and fallow area of 28 household farms averaged 0.62 ha in 1958, 1.37 ha in 1965 and 1.42 ha in 1984. At each point in time, all work was done by hand tools and there was no working livestock and no machinery, though family herds of pigs were raised using cultivated food. Population had grown – some households had vanished, while others had been formed. Significant commercialization had taken place, initially by cultivation of coffee and later by marketing of other produce. While average farm size increased substantially, inequalities were reduced, not increased.

Elsewhere there were major changes in the conditions of tenure. In a rice growing area on the northern coastal plain of Java, Winarto (2004) recorded that all land was operated by a small group of owners before 1960, but substantial immigration after implementation of controlled irrigation had wrought great changes. By 1992 only 32 per cent of farming households were owners, and others got land by sharecropping (the majority), cash rental or other forms of access. There had been much fragmentation and 49 per cent of cultivators used two or three plots while 30 per cent used only a single plot. Through a dense web of work arrangements, many of the landless majority were still able to get access to the produce of the cultivated land.

---

of tenure are also very varied, and it is common for farms to consist of land that is tenured in different ways. Farm expansion often takes the form of renting in, and contraction of renting out. The examples cited here and in chapters 2 and 3 demonstrate the dynamism of land tenure arrangements, along with all other elements of the farm. In the more commercial economies, dynamism extends to transfers both into and out of the capitalist and family modes of production. Academics and journalists may use the dodgy data available to conclude (for example, *The*

Economist, 4 December 2004: 34) that 'farms are generally on their way out, squeezed by high land costs and low profits', but there are still new entrants.

The important considerations in defining the family farm are, first, that the farms be managed by the family group, even if sometimes with significant hired work-forces and, second, that most are smaller units by the standards of the country, region and time. We regard it as much less diagnostically significant that the farm be handed down through generations, although in such cases the 'consubstantiation' of family with farm, discussed by Gray (1998), does give a deeper social meaning to the concept. In terms of labour input, something like half or more should, except in peak periods, be provided by the owning or managing family. The farm may produce all, only some or none of its own subsistence. It may sell all, some or none of its produce. It is the principle of family organization that is most important, and actual statistical boundaries, even if we had the data with which to apply them, would obscure the more essential character of the family farm. Definition has to be both flexible and dynamic.[7]

Notes

1 The much later publication of *Pre-capitalist Economic Formations* (Marx 1965), originally written while *Capital I* was in preparation, makes clear that Marx also considered several different roads from both independent and collective forms of property organization to the separation of a class of capitalists and a class of proletarians.
2 Marxian economics, like classical and neo-classical economics, relies on perfect competition as its working hypothesis. It is also optimistic in that it sees history as a record of material progress. Seers (1979) analysed other parallels and classed Marxian economics as a variant of neo-classical economics, both derived initially from Adam Smith and David Ricardo.
3 We avoid, as far as possible, the arcane terminology derived from Marx and Lenin which was employed by Friedmann and others in the 'new sociology of agriculture' that grew in the 1970s and 1980s, drawing heavily on the Marxist literature. This wave, which paralleled closely similar waves in other social science disciplines, has now died away so that the classics of the period, including Friedmann's papers, are now seldom referred to by modern sociologists (Buttel 2001).
4 He regarded the advanced and progressive differentiation described by Lenin (1899) as illusory, though admitting it as a secondary and minor element. He did not mention Lenin by name, for obvious reasons in the Russia of 1923–25, but referred only to writers 'of a recent period', or of 'the late nineteenth century'.
5 The diagrams used by Chayanov to present this equilibrium under different conditions of production and level of requirements are of a type still likely to be readily familiar to any student of introductory modern economics. However, a more informative and elegant set of diagrams was presented by Ellis (1988a: 102–19).
6 In the State of Missouri, and in some other states in the USA, the law defines a family farm corporation as one in which at least half the voting stock and half the stockholders are members of a related family. At least one stockholder must actively manage the farm. Two-thirds of the corporation's earnings must come from farming (Constance *et al*. 2003: 82).
7 Too late for inclusion in the text, we found a paper by Schmitt (1991) who, after noting that scale economies on family farms have been underestimated, draws attention to the high transaction costs of hired labour, giving a cost advantage to family labour that has widened and is likely to endure.

2 Farming as it was

Turning aside from theory, a few glimpses of farming practice as it was in the first half of the twentieth century can provide a little understanding of how family farming operated before the changes discussed in the following chapters had taken place – before major subsidies to farmers in some developed countries, before large-scale mechanization and chemicalization, before collectivized farming in the communist countries, and before the era of 'economic development' in the developing countries. Two of the glimpses are of conditions in Asia, two in Europe and one in Africa. None are presented as 'typical'. The material is spread over some 40 years, and is presented in order of time, not place. Hand labour was a common feature, working livestock were used in four of the five cases, and the careful planning of labour inputs was required for all. Beyond those remarks, we let the examples speak for themselves.

Kossho, Ibaraki Prefecture, Japan, in 1903–10

In 1910, the eldest son of the leading family in the small village of Kossho, in the Kanto plain north of Tokyo, wrote a novel 'from life' about an impoverished family in his village, for serial publication in a Tokyo newspaper (Nagatsuka 1989). The family rented all its land and was almost permanently in debt, principally to Nagatsuka's own father. The central character, Kanji, was the adopted son-in-law of a family that until recently had owned its house-site, and therefore was probably the remnant of a minor 'branch family' formed from among the bonded servants (*nāgō*) of the only principal family in the small village at some time during the period of major agrarian change between 1600 and 1800 (T.C. Smith 1959). It was a common practice in Japan for a family without male heirs to adopt a son-in-law who would then inherit whatever the family owned, and its debts. His rent, together with debt repayments, was paid in rice. It absorbed more than half the crop of his one small wet-rice field and a small number of dry-crop fields. On these uplands rice was grown, together with the barley that was generally mixed with rice in meals and often formed the principal dish. Most farmers in the village also planted sorghum, soya, sweet potatoes and a range of vegetables. The better-off farmers sold these, but for the poor they were a vital source of food during the months before the main harvest. Some farmers had horses which were used to plough the larger fields and for transport, but the poor had only hoes and mattocks for field preparation and their own backs for carrying loads. The better-off made considerable use of fertilizer, especially expensive dried fish meal, but the poor had only their own night soil, composted with leaves and grass taken from the forest. Even the latter had to be paid for since the once common forest had been parcelled out among private owners in the reforms of the mid-nineteenth century. Farmers helped one another a good deal with field tasks, and exchanged tools as needed. Horses seem to have been used for final puddling of the soil in wet-rice fields on a shared basis. The more substantial labour needed for large jobs, in particular the planting out of rice seedlings, could only be obtained by farmers with sufficient resources to offer the helpers a feast once the job was done.

Nagatsuka's own family employed live-in labour to work their 2.5 ha of productive land. Each man was contracted for a period of years. Although the nature of the contract is not specified, it would seem to be similar to that of the *hōkōnin* of the previous two centuries, with the wage sometimes paid in advance to the head of the contracted servant's family, and sometimes directly to the worker (Smith 1959). The contractual arrangement included food and often also clothing supplied to the worker, both of a higher quality than could be afforded by the independent tenants and their families. This contrast in living standards is made very clear in Nagatsuka's story. Kanji had himself been part of that workforce as a young man, before his marriage, and he remained part of it until his contract was served. The death of his wife's mother gave him inheritance as adopted son-in-law. Kanji and his wife had planned to liquidate debts by contracting out the services of their teenage daughter who, instead, became Kanji's principal support after the early death of his wife in 1903. In a year or two she became his equal partner in field work, as well as his housekeeper and foster-mother to her small brother. In the years 1903–10 Kanji was often employed on a casual basis to open up new fields in the forest, making it possible for him and his family to survive and slowly to improve their situation. In this, Kanji was profiting from his dependent status on his landlord's family, giving him opportunities not available to all the village poor.

Many smaller farmers did paid work on the larger farms during the season, and fell behind with their own programmes in consequence. There was none of the widespread 'put-out' industrial work, common in many Japanese villages, available in Kossho, but 40 km west was the small town of Noda, where there was a thriving soy sauce factory. Young people from Kossho often found work there, with the effect of driving up agricultural wages in the village. From time to time, there was work available on river embankment maintenance some kilometres away, work that paid so poorly that those who did it returned without any net improvement in their financial resources. Although Tokyo was only 70 km distant, none but the wealthy ever went there, and the city did not form part of the work or living space for Kossho's farmers.

The village had one shop, but for services such as a doctor or a blacksmith the villagers had to go to another and presumably larger community. Tensions and jealousies were rife, and gossip could be very wounding. Kanji, who had not been above helping himself to other farmers' crops, was a particular butt of this talk. The community, as such, came together only for major seasonal ceremonies and festivities. But while most of its families were very poor, they were better off than those on the far side of the river, on alluvial land liable to flood, and without nearby forest land. All this is now swept away. Japan experienced a major land reform after World War II, and motorized tillers and other equipment of small-farm scale then became widely available (Francks 2005). Kossho is now embraced within the expanding metropolitan area.

Volokolamsk, Moscow Oblast, Russia, in 1910–13

We encountered Volokolamsk in Chapter 1. It is now the terminus of an outer suburban commuter rail service from Moscow but in 1910–13, when Aleksandr Chayanov (1925, 1966) did important research there, it was wholly rural. Writing about it in the 1920s, he described conditions in that period as though they were still current a decade later. Chayanov's chapter 4, mainly about farm organization in Volokolamsk, provides a careful analysis of the conditions under which family farms of this type and period took decisions on allocation of land, work and income. Chapter 4 was added in the 1925 edition of his book. It did not form a part of the 1923 edition published in German. It therefore post-dates the theoretical formulation introduced in our Chapter 1.

In the last years of the nineteenth century, farming in Volokolamsk had undergone some major changes. Commercial cultivation of flax for a dispersed linen industry was long established in the region between Moscow and the Baltic states, but in the second half of the nineteenth century it expanded greatly to meet the enlarged demand of the then newly developed factory-organized linen industry at Belfast in Ireland. By 1900, Russian flax was by far the main supplier to a world market centred in Belfast. It remained so until 1914.[1] Farmers of Volokolamsk district, with rail connection to the Baltic port of Riga, had taken up these opportunities more vigorously than did those of some other flax growing areas, and thus became one of the more prosperous districts. After about 1890, farmers also began to sow grass with clover to improve the availability of workhorse and cattle feed, and to add nitrogen to the soil before the demanding flax crop. Cultivation of oats for feed diminished and even rye (for bread) was reduced in relative importance (Chayanov 1966).

Even though yield was variable and the price only somewhat less so, flax became the most important crop at Volokolamsk. It provided winter work in pre-sale primary processing to extract the fibre, by retting and scutching to yield the small percentage of the plant that is of commercial value. The standard regional crop rotation progressively gave way to distinctive rotations developed by the farmers themselves to better suit the flax-based economy. Clover always preceded flax in these rotations.

The area of land available was not the primary determinant of land use by middle peasants such as those described. Chayanov was not concerned with differentiation, as Lenin (1899) had been. Land in general was not short because field plots were changed from time to time due to occasional re-partitions of commune land, and there was frequent renting. With only a limited possibility of obtaining hired labour, and a poorly developed system of collaborative labour exchange, it was the family workforce that was the principal determinant of the scale of activity. Crops for sale were determined by market conditions, and the family bundle of needs by its dietary and other requirements, of which a major part was purchased in this village. Beyond that, all other decisions were to a great degree interdependent on one another in the context of strongly seasonal demand for labour.

Work was done with horses or oxen. Reapers, sometimes reaper-binders, and threshing machines were available, together with a variety of field implements, but no tractors as yet. Given the available labour, a first decision was how much this could be spread over the farming year to avoid peak season bottlenecks, especially at harvest time. If the family planted more than it was going to be able to harvest, even with nine- or ten-hour working days during the limited harvesting season, it would be wasting its effort unless it could hire labour. Crops with peak labour requirement arising at the same time were a particular problem, to be avoided as far as possible. Where the peak labour requirements could be separated, the whole task would be much easier to manage. These considerations determined the crop mix and how much land was allocated for each crop.

Associated with the same set of decisions was the very important one regarding the use of horses. One horse could work between two and three hectares of land. The demand was very uneven through the year, and for several months the only use for horses was transport. Farmers on small areas of land found it more advantageous to hire a horse (and workman) by the day than to meet the cost of maintenance. The cost of maintaining horses and cattle was a critical element because in central Russia most farms have only limited natural feed and all livestock have to be stall-fed for several months of the year.

Cattle were essential to provide manure to maintain the fertility of the arable land, although they were sometimes used as draft animals to supplement or even replace the horses. Every aspect of the farm was linked to the constraints in providing feed. Some parts of the farm provided fodder, while most of the arable land consumed livestock produce in the form of manure. In Moscow Oblast, meadow-land was scarce. This was the great value of the sown clover at Volokolamsk, where on representative farms it provided 37 per cent of total livestock feed. The balance came from meadow hay, straw and oats (Chayanov 1966). While even the smallest farms needed to keep one cow to produce manure, farms could expand their livestock production with sown grass or roots, up to the limits at which family and hired labour was available. Some farmers near the railway developed an unusual rotation which included a two-field share of pasture in every year. Though less suited to high yields of flax it enabled these farmers to raise more cows to supply creameries that developed along the railway.

Complications arose from the fragmentation of fields; most farmers had land in 20 or more separate strips. This imposed costs on the use of labour and horses. The distance from farm to field, as much as several kilometres, could involve a doubling of cost between the closest and the remotest plots. More generally, the closer plots received the most attention and manure, but there was great difference between farmers.

Including field work, domestic work, 'crafts and trades', and pre-sale processing of flax, Volokolamsk men used on average 75 per cent of their time productively, and women used 88 per cent. Male youths and adolescent girls did much less. The largest amount of field work went into the production of flax. This was also true of the small number of day labourers (8 per cent of all work days), 43 per cent of whose time went into the crop. Among 'own and short-time workers', about 10 per cent of time was work with livestock, and a further 14 per cent was employed on the meadows and in the clover fields. Together with 6 per cent of time working on the oat crop, this meant that about 30 per cent in total was involved with the farm livestock and their feed (Chayanov 1966). Use of machinery at Volokolamsk is not separately analysed. Its value, however, totalled only a little less than in a wheat region of southern Russia and was well above that in other north-Russian areas. Most Russian farmers of the early twentieth century delivered their own produce to market, but intermediate dealers were involved in the longer market chains, such as that of flax. Chayanov (1966) provided one of the first discussions of these 'supply chains' in the literature, with some valuable diagrams. Some of the dealers were undoubtedly able to enmesh suppliers in very onerous credit arrangements.

The Nupe Emirate, Nigeria, in 1934–36

Dr Fred Nadel spent two long periods in the Nupe Emirate in the middle belt of Nigeria in 1934–36. He wrote a rather unique anthropological account of the society, politics and economy of half a million people (Nadel 1942). The Nupe, who trace their history as a nation back to the fourteenth century, live to the north of the more numerous Yoruba. They are separated from the still more numerous Hausa of the north by a number of peoples most of whom had in common that they were conquered by Muslim Hausa or Fulani in the early nineteenth century. Advancing further south, the Fulani conquered Nupe in the mid-nineteenth century.[2] Then, in the course of their advance northward from the coast, the British conquered Nupe in 1897. In the mid-1930s, this still partly pagan, partly Muslim people of varied origin had experienced less than 40 years of colonial rule. The effects of Islam were varied; Koranic education was all that most rural boys got in that period (girls getting none), and while the end of Ramadan feast (Id-El-Fitr) had become the principal festival of the year, there was little observance of the fast itself.

British money replaced cowrie shell currency, but Nupe had experienced special-ization of labour, and monetized trading, for centuries. The British introduced a general tax that replaced the local taxes of the past, the slaves of the principal nobles became sharecroppers on the same land, and new avenues to power and relative affluence emerged through the 'Native Administration', but village government was little changed. Although cotton and groundnuts were grown principally for sale, most crops were used both for household purposes and trade. Complex crop rotations were practised, but almost all land was rested for long fallow periods after each few years of cropping. Cattle were fewer than in pre-colonial times. Northern herders, driving their cattle south to the Yoruba markets, sold some to the Nupe butchers.[3]

Land was held under a hierarchy of titles: by the king, by the village chief or by the lineage and household groups, and by individuals. Most individuals held some pieces of their own land and worked with only their sons, but household groups (*efakó*), working together, were still the main effective holders and working groups. The agricultural year began in April, and as soon as the first significant rains fell farmers would begin the task of tilling the land into ridges or into mounds for yams. In this middle zone, both the southern root crops and the northern grains (millet, sorghum and some maize) and groundnuts were cultivated. In 1934, the price of groundnuts had fallen so low that much of the crop was left in the ground and some farmers shifted their inputs to red peppers which maintained a better sale price.

The agricultural year involved complex scheduling of tasks, with significant peaks for tilling, and for the first and second weeding. At the year's end the cotton harvest prevented many farmers from preparing and seeding new yam mounds to be ready for the next year's growing season. The peaks in labour demand were the occasion for group labour. Some of it was arranged casually between neighbours. Larger groups were organized through the age-grade system, accompanied and encouraged by hired drummers and musicians and reimbursed with food and sorghum beer. These groups always served the village chiefs first, but were available to all who could afford them, and used in such work as house building as well as in farming. Agriculture was heavy work, all done by hand, and farmers took considerable pride in the straightness of their ridges and rows of mounds. 'The skill and keenness of a farmer is judged by the clean look of his farm, with all weeds, leaves and brushwood cleared away. An overgrown, untidy farm invariably elicits the comment: *za-gá de kokari à*, "that man makes no effort"' (Nadel 1942: 208).

Land was not yet short in Nupe, except in areas close to Bida city where a form of land rental, borrowing with a payment, had already come into existence. In the less closely settled areas, new independent families could still get land from the head of the kinship group, the community or its chief. Islamic law had not displaced indigenous practice in the matter of land allocation and inheritance. Women had no rights in land, and all farm work was done by men. Women cooked the food, and did most of the work associated with certain trees valuable for their fruits, especially the shea nut and kidney-bean trees. Ownership rights over these and some other trees were, however, specifically vested in the chief, a conservationist provision also found in northern Ghana to this day (Gyasi *et al.* 2003; Gyasi and Asante 2005). The main and very important role of women was in the marketing of farm produce, to which we return in a wider African context in Chapter 5.

At the time of Nadel's field work the colonial opening of the economy to foreign trade had its most damaging effects on local industry, which had been highly organized both on an individual and household group (*efakó*) basis. By the 1930s, the indigenous glass industry was almost dead, weaving was facing serious competition from imported cloth and yarn, and even the blacksmithing industry, closely integrated into agriculture, was severely depressed through imports of bush knives and of blocks of raw iron which reduced the prices obtained for the iron smelted from local lateritic iron ores. Blacksmithing remained an important activity both in the capital and in all larger villages. The actual making of agricultural tools was conducted mainly on an *efakó* basis, and some operations still smelted their own iron from the local ores with charcoal of their own manufacture. Repairing work was done mainly by individuals, whether or not members of an *efakó* group.

Other accounts of African farming in the 1930s tell a similar story. Many of the agronomic findings of this work are brought together by Allan (1965), and we have ourselves followed and analysed the whole large literature on the system of the Bemba of Zambia, which began in the same period as Nadel's work (Brookfield 2001).

Farmers in Yunnan and north China, in 1937–49

There is a very large literature on Chinese agricultural and agrarian history. Netting (1993) wrote one of his best chapters from a survey of this material. Our use of the literature is highly selective. The Japanese invasion in 1937, preceded already by an occupation of areas between Beijing and the Manchurian border, had two notable effects for the study of China's farmers and agrarian society. Anthropologists at an institution that later became part of Beijing University had already begun rural studies in China. In 1937 and 1938 most of the staff and students set up new quarters in Chengdu (Sichuan) and Kunming (Yunnan), and from the latter carried out studies in contrasted Yunnan villages between 1939 and 1942 (Fei and Chang 1945). Meanwhile, through the curious sponsorship of the South Manchurian Railway Company, Japanese social scientists (with a few Chinese colleagues) surveyed the land and society of 33 villages on the north China plain in Hebei and north Shandong. The purpose was, of course, to gather information to assist Japanese military-colonial administration, but before we throw up our hands in horror about this, we should recall that most African anthropology, including Nadel's work, started in this way with the direct sponsorship of the colonial authorities.

Villages in Yunnan, in 1938–43

The three Yunnan villages studied by Fei and Chang (1945) were strongly contrasted. The objective was to explore social relations as expressed in land use and tenure, and rural industry. In this mountainous region with only limited agricultural land, local as opposed to general population densities were very high. Units of land and labour were expressed in the *kung*, meaning both the area of land that could be worked by one worker in a single day and – with greater imprecision – the measure of a day's work. The size of a *kung* varied according to ecological conditions, but was generally not more than one-third of the national *mu*, then still somewhat variable regionally but now standardized at one-fifteenth of a hectare. In all three villages land was very unequally distributed, but a family that held less than 15 *kung* (about 1.05 ha on average) could not make a living entirely from its own land. In each village this was at least one-third of the population. We discuss two of the three villages below.

The first village had no absentee landlordism. It is in a fertile and relatively low-lying basin about 100 km west of Kunming. Although close to the then new Burma road (now the old highway between Kunming and western Yunnan), it had few commercial relations beyond its own valley, and there was only minor and supplementary non-agricultural employment. Everything depended on farming, principally for rice and a winter crop of broad beans (*Vicia faba*) grown on only part of the land. In 1938–39 there were 84 farms, the largest of which covered about 1.67 ha, and 38 households had no land. Fei and Chang calculated that farm families had sufficient labour for all work except in the March–May season when land was being prepared for the rice crop. Even then most farmers could manage by using labour exchange among a group, who staggered work. This was widely practised.

However, a whole group of better-off male landholders preferred to undertake no farm work, although their wives and daughters did so. They employed the landless and some migrants as either part-time or full-time labourers. Yet land was not becoming concentrated in fewer hands; partible inheritance, dividing land between surviving male heirs, could reduce a 'leisure class' family to the status of poor farmers in a single generation. The villagers rented out very little land of their own, but rented in significant land held collectively by lineage groups and temples, and from some individuals living in other communities. While substantial areas were held collectively in all villages, none was worked collectively at this time. Tenants on land of their own lineage, indeed on most collective land, enjoyed very secure tenure. The distribution of operated land was much less unequal than that of owned land.

A second village, some way south of Kunming and close to an old town, was different in that it produced vegetables for sale as well as rice, and there was a domestic textile industry employing women in weaving cotton cloth. Merchants in the town controlled this industry and, since the demise of local spinning several decades before 1940, they supplied the machine-spun thread and then bought the cloth. The women received a de facto wage which was less than the value of the food they themselves consumed. Yet the income was a valuable supplement to the resources of a poor household. In this village, where even the middle group of landholders owned on average only half a hectare of rice land, a high proportion of land in use was rented, including all the intensively cultivated land under year-round vegetable gardens. Most of the landowners lived in the nearby walled town and carried on business there.

In the three villages about two-thirds of the households had either very little land or were landless. Many of them rented small plots, but had to depend on working for the richer farmers for most or all of their income. Men working as year-round labourers would never make enough money to marry. Few of the rich had inherited their relatively large holdings. Most made their money outside agriculture. In this region of Yunnan, some had made money through the illegal and risky business of smuggling opium from Burma, where it was still grown, for sale to the large number of addicts still surviving from the days when Yunnan was a major opium producer. They could and probably did purchase opium also from minority groups in the mountains of southwestern Yunnan itself, where the crop was still cultivated until the 1950s (Yin Shaoting 2001). In Yunnan, wealth could be made by renting out land, in business, trade, and even in the army and government, but not by hard work in agriculture.

The north China plain under Japanese occupation

In the north China plain (Hebei and northern Shandong), historian Phillip Huang (1985) was able to rely on a much richer vein of information than that obtained by Fei and Chang in their underfunded field inquiries in Yunnan. This included not only the 33 village data obtained by the Japanese, but several other surveys of the inter-war period and a good deal of historical data that he researched in Beijing. The whole region was more densely occupied than mountainous Yunnan, and had been for more than two millennia. Although commercialization was patchy, it was a major element in the economy and had become more so since cotton cultivation expanded in the early twentieth century, offering higher returns than the older cash crops of wheat, soya and rapeseed. Cotton served the growing factory- and the household-based textile industry. Farmers in the plain in the 1930s mainly cultivated their land on a three-crops-in-two-years rotation. A spring-sown crop of sorghum, sometimes maize, was followed by an autumn crop of wheat to be harvested in the spring, and then by a summer crop (commonly soya), then fallow. Millet, barley, sweet potatoes and cabbage were also grown. Cotton could be fitted into this risk-reducing and labour-smoothing rotation only by using fields with a separate cotton/fallow rotation. Cotton's long growing season eliminated both the summer crop and the autumn-sown wheat. Farmers with three or more fields developed a complex rotation in which both cotton and the older cash crops could be handled.

With a frost-free season of only six or seven months, farmers had to cope with peak work seasons in both spring and autumn, especially the latter, when summer crops had to be harvested and the land prepared and planted to the autumn-sown wheat within the space of only about six weeks. Cotton, therefore, was a high-gain but high-risk alternative, and most farmers only planted it on a part of their land, retaining the rest under rotation. The main exceptions were the land-poor, who had to choose between retaining the security of the food crop rotation (though they generally sold the wheat and bought additional and much cheaper sorghum), and putting land only under a cotton/fallow rotation. They might do very well, but a spring drought could wipe out the cotton crop and their livelihoods with it.

Unlike the situation reported by Fei and Chang in Yunnan, it was possible to accumulate land by successful cash-cropping in north China. Cotton and wheat incomes were used to buy land from small farmers who had gone under by loss of a crop or other misfortune. There was thus social mobility both upward and downward. Smallholder farmers, who rented land at rates amounting to up to half their crop, were at constant risk of descending into the landless or near-landless class. Then they would have to rely almost entirely on wage labour, either seasonal or, more terminally for their own independence, year-round live-in employment on the land of large landholders. As in Yunnan, this often meant that they were unable to find the means to marry, especially as there was no paid agricultural employment for women. Some of the characters in Liu Ching's (1964) novel 'from life' about a village in central Shanxi in the 1950s had been year-labourers who had entered into more informal relationships with women. One or two were able to marry and set up house in their home village after receiving land in the land reform. By the mid-1930s nearly half the households on the north China plain were farming less than 10 *mu* (0.67 ha), when one hectare was about the minimum size for a viable family farm in this region. Although some took the risk of specialization in an attempt to rise, most could adopt no more than a survival strategy in which wage labour had to provide an essential part of their livelihood.

There was one other source of support, as in the second Yunnan village. Wives bought machine-spun cotton yarn from merchants who then bought the cloth they wove. If women owned one of the 'iron-geared' looms introduced in the 1920s they were able to produce cloth at a less-than-subsistence 'wage' they had to accept in order to compete with the factory-woven cloth. Hand-woven cloth still supplied more than half the Chinese domestic market for cloth in the mid-1930s. Many of the plain's households were kept intact and alive by the interlocking of uneconomic small farming with female handicraft work at starvation pay, and the rather more rewarding day labour for those men who could get enough of it. Already many of the year-labourers were drawn from the 'floating' population of landless migrants who were not members of the local community, not always even from the same province. It was to avoid joining this entirely proletarian population that poor households heavily used their own labour, and endured exploitation by landlords, employers and merchants to keep a foothold in something they could call their own.

The other face of social differentiation was the accumulation of substantial landholdings by those who were successful. Bearing in mind that the accumulated land was almost never in one piece and was usually scattered over quite a large area in many small plots, those who had accumulated about three hectares of land had much more than they could use with family labour, but not really enough to make rental income look like the road to wealth. In the north China plain, one farmer could work significantly more than a hectare of land with the aid only of cooperative labour among relatives and neighbours and some hired help at peak times, and large households with several working males might jointly work farms as large as three hectares, without year-labourers. Farms with more than about five hectares always hired labour and were likely to employ most of the labour put into their land, though this hardly made them 'capitalists'. These 'rich peasants' of the pre-revolutionary period gave way upward to what Huang and others called 'managerial farmers', a few per cent of the total landholding population who might have worked 10 per cent of the land with hired and closely supervised workforces. Mostly, they operated areas between about 6 and 13 ha, beyond which close management became difficult. The biggest landholdings were rented out. The managerial farmers were able to exploit their harder worked labour force more effectively, working 252 days in the year rather than the average 180 days for middle farmers. Huang examined all technical reasons why the managerial farmers might use labour more efficiently and he found none that really provided an answer. They did not use livestock or capital more intensively than other farmers. Teamwork in wheat sowing involved four or eight workers in one or two groups, ploughing or hoeing at the front, then seeding, spreading composted fertilizer and covering the soil last. This was no managerial peculiarity but was practised through cooperative labour by all farmers. On a larger scale, and with much more use of labour, the same procedure was followed on a collective farm in Shanxi in 1971 (Hinton 1983). Managerial farmers' input average was only 11 man-days on each *mu* (0.067 ha), in contrast to the village average of 16–19. They got the same yield.

Cold Aston, Gloucestershire, England, in 1942–43

The final short glimpse is of only a single farm. It does not rely on the literature, but on Brookfield's memory. In the wartime years 1942–44 he spent three summers working on English farms, two of them at Cold Aston on the Cotswold hills of Gloucestershire, a rolling upland limestone area that had, in former times, been an important producer of long-wool sheep for the pre-industrial revolution textile industry. The farm was on the dipslope of the inferior (i.e. lower) oolite, described disparagingly by a neighbouring farmer as being 'poor owlite' that yielded only modest returns from its soils. The sheep had long since gone. Their place was taken by cattle on mixed farms that had survived by shifting emphasis from wheat to cattle, from the arable to the ley in the midland system, during the long lean years for British farming between the 1870s and the 1930s.

In the great wartime 'plough-up' campaign of the early 1940s they returned to wheat production which was more demanding of labour. The farm described was a little over 40 ha in area, run by a tenant farmer who employed two middle-aged labourers and, in 1943, a tuneful Italian prisoner of war. For the summer seasons this workforce was supplemented by two schoolboys. Three horses and two carts, one large and one small, scythes, sickles, hoes, pitchforks, a plough, a harrow, a reaper-binder and other ageing tools constituted the equipment. A few cattle were still kept. The first job the two town-boys were given was to clean out the cobblestone-floored cowshed of the accumulated dung and urine-soaked straw bedding, and stack it outside to weather, so that it could be spread as manure before the next crop was sown. They were further instructed in rural realities by a range of other jobs chosen for this purpose.

Later, they followed the reaper-binder around the two wheat fields collecting the sheaves it threw out into stooks to dry, then a day or two later forking these onto a cart to be taken to the farm, where they were stacked. While the reaping was in progress, the farmer's elderly father shot at the rabbits which bolted for safety out of the diminishing block of standing grain in the centre of the field. The boys were instructed in scything lodged wheat in a deep soiled part of the main field, but with poor results. When threshing time came, a tractor-drawn hired threshing machine came into the farmyard, and the work of feeding the machine went on through two hot days, during which a flagon of the local rough cider circulated among the workforce. This was a small farm, with none of the new machinery that was arriving in more intensive arable areas, which in those wartime years also received 'land-girls' of the Women's Land Army. The women were recruited to replace absent men on a more continuing basis than teenage boys' vacation labour could provide. The farm made limited use of the chemical fertilizers that were already coming into general use, though at fairly low levels of application until after 1950.

Yet farms like this, and its workers, were relatively well off by the general standards of agricultural western Europe at the end of the 1930s (Lamartine-Yates 1940). The old farmhouse was comfortably furnished, if the labourers' cottages were not. The drystone field walls were in good condition, and the field edges and pastures were relatively free of weeds. Although by 1942 there had already been considerable improvement on conditions in the 1930s, these had been very bad in other areas of the western midlands of England. Writing generally of Britain in the 1930s, Stamp and Beaver (1941: 145) described big areas infested with weeds, and argued that under better economic conditions 'the soil of the British Isles could be made to produce 50 per cent more, possibly 100 per cent more, than it does at the present time'. The doubling was achieved between 1939 and 1944. With guaranteed prices, though still confronted by the vagaries of the weather, farmers could plan their activities in ways not possible in the preceding half-century. As it turned out, these favourable conditions continued through the rest of the twentieth century.

Notes

1   As in all other parts of European Russia that relied on trade with western Europe, the trade was virtually halted in 1914 with the outbreak of World War I, first by closure of the Baltic route and then of the Bosporus. Only circuitous outlets via the White Sea, Murmansk or Vladivostok remained. By 1919 the German navy no longer imposed any obstacle in the Baltic, but Riga (in newly independent Latvia) was no longer a Russian port, and the chaotic situation in post-revolutionary Russia prevented any re-establishment of the pre-1914 trade pattern. It never recovered.

2   Fulani incursion took much the same basic form right across West Africa. Fulani were, and some still are, semi-nomadic pastoralists, whose migrations took them southward during the dry season into the lands of agriculturalists whose fields were manured by the cattle. Fulani adopted Islam at fairly early dates, and a minority became teachers among the farming people. Some Fulani settled and became arable farmers and others concentrated into towns. At different times in the eighteenth and nineteenth centuries they embarked on conquering wars establishing Fulani-ruled states, sometimes driving out the displaced pagans, sometimes ruling and converting them, and sometimes enslaving them. In Nigeria in particular, they also conquered and ruled some already Islamic people. The tsetse-fly (infecting their cattle with trypanosomiases), as much as armed opposition, determined the limits of their southward advance except where they settled wholly to non-pastoral lifestyles.

3   A subsequent increase in cattle numbers in the whole central belt of Nigeria, especially since the 1970s, has both transformed the economy and created the means of restoring land without the long fallow periods which were normal in the 1930s (Tiffen 2006).

3 Setting up the farm

Accessing land and water

How do family farms obtain the means with which to operate in relation to their social and economic environment? This topic is addressed in this chapter and the next. Our time focus is as close to the present day as we can get, but we have to rely on literature that goes back as far as the mid-twentieth century. It would be better to say of the sources, including that of Brookfield's own field work, that we are writing of the 'ethnographic present'. There is great dynamism in land tenure arrangements, and to understand this and extract the underlying principles we need to consider changes that have taken place in the past. The external relations of the farm inevitably intrude, but are mainly discussed in subsequent chapters. In family farming, as Polanyi (1944) pointed out, neither land nor labour are true commodities, and this is among the reasons why family farming has obstinately failed to vanish as predicted. To better follow the manner in which most family farming has become commercialized, we discuss the non-market aspects of access to land before we turn to the commercial ways of acquiring it. The same approach is used in chapters 4 and 5. We generalize, but recognize that every farm is different and every farming region unlike all others. The related question of water management is also discussed in this chapter.

Accessing and holding onto land

Men

Over much of the world, most individually owned farmed land belongs to men. The most common way for a man to acquire land is through inheritance or to be given it sooner, normally by his father but in some matrilineal societies by his mother's brother. Many farming families strongly wish the farm to continue within the family, and in Europe some farms, even tenant farms, have continued within the family for as long as two centuries. Where partible inheritance operates, sons may receive only a part share in the patrimony. If one son receives the whole farm, he is obliged to pay out his coheritors in some way, in Europe often taking out substantial loans to do this. In many developing countries title to land is not individual, and what a young man receives is not a piece of land, but the right to cultivate land that is owned collectively by a kinship or kin-like group to which he belongs. In his long unpublished work of 1857–58, commonly known as the *Formen*,[1] Marx (1965: 68) regarded this latter as the 'earliest form of landed property', and discussed it at length. Whether or not the earliest, it still survives in many parts of the world.

While it is usually the eldest son who receives all or the major part of the parental farm, there are many instances in which the youngest son receives it, remaining with his parents until they are aged, and having the responsibility for care and meeting all ceremonial costs for which the father was responsible and the funeral costs on his parents' death. Since the elder siblings have meantime acquired livelihood in other ways, this system often assures continuity of the family farm.

Where there is dispute, as happens in countries that have been wracked by civil war or where there is no surviving record of land allocation, claim to land is enhanced by improvements and especially by the planting of economic trees (Unruh 2006). Many landlords in developing countries refuse to allow tenants to plant trees, from concern that they might later use them as the basis for a land claim. Claims on the basis of prior occupation, with landscape evidence as the main support for otherwise verbal statements, are still important in many countries.

Women

Women inherit land in many parts of the world, and do so as coheritors wherever partible inheritance among all children is the rule, as in much of southeastern Asia. Kinship is widely recognized on a bilateral basis through the nuclear family, and a new farming family puts together its inheritance or prospective inheritance from both sides. Commonly, the families live with the wife's parents initially, setting up their own household when children arrive. There are a few cases in which land goes to women exclusively and remains with them. The best known case is the Minangkabau people of central Sumatra and part of the Malay peninsula. Matrilineages own the land; effective control resides with the senior woman in each sublineage, within which households form with different relationships to the senior woman. These are composed of core families, client families whose (female) ancestors were not members of the original core community, and servant families who, while now equal in status, do not possess land. Men depend on their mothers for use rights, which never become permanent property. In one Sumatran community more than three-quarters of the land belongs to households of the core matrilineage (Blackwood 1997). Even with the much more substantial modern changes that have taken place in Malaysia, the formal title to over two-thirds of all land in some Minangkabau villages is held by women (Brookfield *et al.* 1991).

There are many societies in Asia and Africa where few women inherit land. In strongly patrilineal societies, the male members of a lineage or society constitute an ongoing corporation and marriages are normally patrilocal, so that land given to a daughter would pass out of the lineage on her marriage. The alternative is to adopt the son-in-law as a member of the lineage. Although widows are often left with at least some of the land after their husbands die, many have to depend on male relatives for access to land they have in part or even wholly managed for years. Even in areas where land rights are not rigid and depend on interpersonal accommodation among families and their members, women often find it difficult to secure access to the land they want to use. Frequently, the complex bundle of rights in resources is disaggregated so that specific responsibility for certain resources belongs to women, whether or not with formal recognition (Meinzen-Dick *et al.* 1997). This applies particularly with rights to trees, where specific parts and specific stages of growth may provide 'gendered spaces' for women (Rochelau and Edmunds 1997).[2]

Operation of land by women, in contrast, is very common. In general, it is only in very modern times that ownership rights arising from operation have begun to be legally recognized. Even in Europe and North America, only a fraction of all farms are legally held by women, most of whom are unmarried, divorced or widows. Rarely are farms worked by a married couple held in the wife's sole name. However, growing numbers of farms are held in legal partnership between husband and wife, and sometimes also with an intended heritor. Partnership, formerly mainly among men, is an old device but becoming increasingly popular. Prugl (2004) comments that German lawyers caution men about partnerships with their wives in case of problems on divorce.

Where not mandated by law or custom, inheritance is a highly personal affair, a power over the young which older people tend to hold to themselves. Transfer of property does not occur only at death; it may be a continuous process. The heritor often works on his father's or uncle's land for a protracted period and also assumes full responsibility for the care of the old people. Where there is more than one child, inheritance is rarely a certainty. An elder son may be displaced by a younger, or even by an in-marrying sister's husband who demonstrates greater aptitude, or has fathered a favourite grandchild.

Land and the organization of society

A piece of land may change hands in many other ways than by inheritance. For farmers, all productive activities are earthbound, and when people organize themselves according to kinship and descent – the principles they mainly employ for inheritance – they are also organizing their management of land. There is a constantly recurring need to adapt rural social structures to reality on the ground. People may think of themselves in terms of their genealogies and their marriages, but they also form local groups with unrelated neighbours important to them in everyday activity. The simple fact that land stays put while people move and form new relationships underlies the dynamic relation of landholding to social organization. Close examination often discloses that what actually happens is not always according to the supposed rules. A particularly detailed study of a village in the dry zone of Sri Lanka, where there are written records since the 1880s, led its author to conclude that:

> the kinship system is not "a thing in itself" but rather a way of thinking about rights and usages with respect to land. The land is fixed; the people change.

Ownership is determined not by simple rules; it evolves and fragments by the processes of gift, sale, and inheritance, but all the time the land is in the same place and the "owners" must adjust their relationships to conform to this inescapable fact.

(Leach 1961: 146)

Provocatively, Leach (1961: 305) went on to write that 'kinship systems have no "reality" at all except in relation to land and property'. At the time, this was not an enormously popular view in a profession that had spent a generation building up elaborate constructs about society in which the basis was the kinship system. Brookfield was still feeling his way into his involvement in anthropology at the beginning of the 1960s, and found that he had to tread carefully in this minefield. His caution continued through publication of work fundamentally concerned with how land was acquired and occupied in a densely peopled region with notionally patrilineal social structure, but with substantial absorption of matrikin and affines (Brookfield and Brown 1963). This work formed part of a growing – and sometimes heated – debate in which an increasing number of writers cast doubt on the validity of a purely structural approach to the understanding of society, and placed greater emphasis on locality as the organizing principle.

Borrowing and giving land

The simple and often informal borrowing of land – between neighbours in particular – is still sometimes practised in developed countries, but it takes place more widely in systems where land-rotational cultivation (shifting cultivation) is practised. There are many reasons, both practical and social, why farmers should want to work land in proximity to their neighbours and relatives. They may borrow a plot of fallow land ready for cultivation that the owner does not wish to utilize. With a long fallow cycle this means the owner must give up the land until the next fallow cycle is complete. Yet this happens frequently where the pressure on land resources is low enough to permit the sacrifice. Borrowing requires no rent, but does create an obligation to return the favour in the same way or by other means.

Borrowing may be an integral part of social organization in societies where land rights are individualized only to the degree that individuals identify pieces of land as their own within a formal group territorial claim that is based notionally on kinship. Individuals have personal networks that distinguish between the people with whom they have close interaction and more distant relations, and those with only tenuous connection. In northern Java, the holder of an adjacent block of wet-rice land can be the closest of all non-family to a farmer (Winarto 2004). Using a specific example derived from Lawrence (1955), Brookfield with Hart (1971) used set-analysis to show how local and formal groups overlap in the choice of partners to work adjacent land. This means that they clear and prepare the land together, and trust one another throughout all stages of the cropping cycle. To do this, individuals may have to borrow land from a formal group other than their own, but who are members of their personal network and mostly from the local group. More detail on land borrowing in shifting cultivation societies is discussed in Brookfield (2001).

The distinctions are not sharp between loan and gift of land, or even rental. While most giving takes place amongst close kin, some land is given to people to whom donors are indebted, or to recipients whom they wish to indebt. Where there are registering authorities that tax formal transfers, a lot of land is 'gifted' to avoid or reduce such charges, even though specific payment has been made. Transferring use of land in these ways can impact on the way a formal group itself is perceived. Formal groups are necessary for many social purposes, of which one of the most important is providing access to land for their members. But formal groups operate within local groups, which are the effective units of joint activity, and in many cases membership in a local group is the only effective guarantee of access to resources. In a majority of the world's local groups there is no cadastre, and no formal register of population, so it is quite easy for the genealogies which validate formal group membership to become modified, even in the space of one generation, to legitimize the presence of long-term borrowers.

Some long-term 'borrowers' become 'de-facto tenants', recipients of land who make no payment as such, beyond some work and support for the landowner. This was the case in the Chimbu area of Papua New Guinea.[3] Between the informality of these arrangements and the formality of a written and signed contract there lies a very wide range. But in a great many instances the authority of important landowners ensures them respect and service, if not also fear of their displeasure, even where substantial rent is paid.

Much the same happens in Africa and it is highly likely that it happened in historical Europe and Asia. China would seem to be an exception. The patrilineal clan or lineage of southeastern China was an ongoing corporation which not only legitimated the rights of its members, but also held land of its own that was rented out to members at a modest charge (Siu and Faure 1995; Yang 1959a, 1959b). Since the lineages often included people who lived outside the principal village, they held land at a distance which could be rented to others. Alliances of villages formed through the lineages. Although the lineage might have only a few literate members, they had an important role in recording genealogies. When large lineages – some numbered in the thousands – divided into more manageable groups the genealogies were maintained within the partitions and this was when errors could creep in. The lineages emerged in more than one way between the fifteenth and eighteenth centuries, and before genealogies were written down they seem to have operated in a more flexible manner by incorporating newcomers (Faure 1989). In a formerly more fluid situation, they established a structure of local government which served both their own purposes and those of government in sponsoring them. This led to rigidity in land allocation which exacerbated class distinctions within and between rural communities. They were extinguished as landholding groups in 1950, but re-emerged strongly after collectivization ended (Chapter 7) (Liu and Murphy 2006).

Commercial transactions in land

Farms can readily enough be created or enlarged by purchase if the farmer has enough money or credit. 'Sale' is a term with many shades of meaning and, notwithstanding the absence of any formal land market, land was bought and sold in classical times. It is conventional to regard the holding of a written title as a necessary condition for the sale of land, but it was not always the case, and not everywhere even now. The buying and selling of land preceded capitalism by a long way. In the nineteenth century it was handed out as an (almost) free gift in the new settlement of the western USA and Canada and at once began to be sold to new buyers. In Britain, early modern land reorganization created a landowning class that owned three-quarters of the land but worked little of it; most farmers in the nineteenth century were tenants who worked their land with own or hired labour. Declining land values, and a shift of wealth from rural into urban land and into company shares, enabled so many leaseholders to buy out their leases after 1900 (in fact, principally after 1918) that the proportion of owner-occupiers among all farmers increased from a little over 10 per cent in 1900 to 36 per cent by 1927 (Hobsbawm 1968). This major land reform required no active intervention by government, but some great landholders remained stubborn. Leaseholders continued to campaign for the legal right to buy their land right through the twentieth century, winning it in Scotland only after self-government in the early twenty-first century. The prestige that comes from owning rural land has been democratized, but great landed estates have not died out in western Europe, in Britain in particular. In the USA it was once feared that concentration of land ownership would create a landowning aristocracy but this did not happen; massive concentration has taken place but mainly into corporate hands (Vogeler 1981).

Buying and selling of land has been substantial in the developing countries as well. In northern India there was substantial sale of land during the Green Revolution period, as some farmers with little land sold it to concentrate on the new employment and entrepreneurship opportunities (Leaf 1984). Lower caste farmers bought land from the Brahmin farmers who had previously held and probably rented out most of it – four-fifths of it at one Uttar Pradesh village (Wiser 1978). In a rice growing community in northern Java, developed under centralized irrigation since the 1960s, several waves of immigrants have bought land from the original holders. They sold it for a variety of reasons, but principally to build new houses or to perform the pilgrimage to Mecca (Winarto 2004). The total area of land bought and sold between developing country farmers and land speculators is very substantial although there are few even 'educated guesses' as to how much has been transferred in this way. Even where the documentation of land transfer is a legal requirement, and where taxes are paid on such transfers, record keeping is often imperfect.

Markets in land can and do develop even where they are specifically disallowed by law. From German times in Tanzania, for example, all land belonged to the state and individual occupation is a use-right, not ownership. Over time a growing amount of land has been transferred between individuals for sums rationalized as payment for 'unexhausted improvements', which can include the cost of its initial clearance. By the 1990s, so much land had been bought and sold in this way that the land market was socially legitimated, and transfers of holdings were normally registered to ensure the security of the new owner. Even though it is said that 'in Tanzania we do not sell land', in practice by 2000 there was a market in private rights to land as effective as in any other country (Daley 2005). Similar transitions have now taken place in many countries, so that the absence of effective individual ownership from arable land is now becoming the exception. Government land titling programmes add legal force to this trend, and are strongly pressed by the international financial institutions for doctrinaire reasons, but in many cases these costly programmes can do little more than ratify what has already taken place.

Land and credit

Short-term credit is an almost unavoidable requirement of farming, given the need for expenses early in each season long before any crop can be harvested. Only the relatively affluent can get by on their own. Quite apart from farming expenses, farm families also have to meet regular expenses, for example for children's schooling, and the more costly life-crisis events – sickness, birth of a child, marriage and death. Within rural communities across the world those with resources have given loans to those without, commonly on repayment terms that constitute high rates of interest. For larger loans security is required and this is most commonly the land. Some loans are not repaid, and land may be transferred. Mortgaging of land can be a dangerous business in times of rising costs and unstable prices, and many developed country farmers lost land in this way during the agricultural downturn of the last 20 years of the twentieth century. Such losses are not limited to the developed countries; Box 3.1 provides a striking instance from Malaysia.

The close link between credit, usury and land accumulation has been remarked on by many writers, not least Marx and Lenin. Widespread awareness of the need for credit, and its perils, led to some important initiatives in nineteenth-century Germany, where Friederich Raiffeisen pioneered rural credit unions or cooperative banks, which subsequently stimulated incipient rural production and marketing cooperatives. Rural banks on Raiffeisen principles continue to exist all over central and eastern Europe, and a similar approach to micro-credit has been adopted in a number of developing countries since the 1970s. Larger and longer-term loans are nowadays mainly through banks, including the agricultural banks that have been established by government in many countries.

Renting and sharecropping land

Those who do not inherit land or receive it by transfer among living families, who do not get it by grant from a tribal chief, government agency or a powerful patron, who cannot find a lender and who cannot afford to buy it, normally have to enter the rental market. Renting land is central to the achievement of flexibility in land tenure. It is a principal means by which a farm can be set up, or enlarged, and by

---

**BOX 3.1 HOW MALAYSIA'S 'HAJI BROOM'
ACQUIRED LAND**

James Scott's (1985: 15–18) 'Haji Broom', in the Kedah rice plain of
Malaysia, did little more than a great many others who lent money for land,
but used some unusual tricks. Operating on the basis of promised sale (*jual
janji* in Malay), widely used in Southeast Asia, he would lend money some
six months ahead of the harvest against a written contract, under which he
would take charge of the title deed of the pledged land, always of greater
value than the loan. If the borrower repaid in time, the title deed would be
returned, but 'Haji Broom' made a practice of going into hiding a few days
before the repayment date and then, without waiting for the borrower to find
him, he would go to the government office to register his 'land purchase'. In
something like this way, many families lost land to him, and by the time of
his death he was reputed to have accumulated over 170 ha of land, in an area
where a normal 'rich farmer' might have five or six hectares only. Developed
country banks can sometimes be just as iniquitous with their borrowers. In
the USA Vogeler (1981) offers a striking example.

---

which older farmers can retain income from land they are no longer able or want to
use for production. The speed with which rental markets have developed in regions
where collectivization has been abandoned has astonished many observers, and it
has been very instrumental in enhancing productivity, by enabling those best
endowed by age, resources and skills to gain access to land. It has also speedily
facilitated the re-emergence of inequalities (Haroon Akram-Lodhi 2005). There is
a great range in the type of tenancy arrangement.

Commonly, it is argued that full title to land, even on mortgage, is a step higher
up the 'ladder' than renting land, and there is no doubt that most farmers prefer it.
But while tenancy means paying rent, many farmers pay substantial interest on the
loans they take out for ordinary operating expenses, as well as for land purchase and
capital improvements. Writing of western Europe in the 1930s, Lamartine-Yates
(1940: 524) wrote that 'the large majority of peasant proprietors have to pay in
interest as much, or nearly as much, as those who rent their farms'. The farm crises
of the 1930s, 1980s and 1990s in the developed countries hit the indebted landowners most heavily. On the other hand, it was the overwhelmingly tenant farmers
of the commercial rice producing areas of Vietnam and Burma (Myanmar) who
were the principal victims of the collapse in rice prices in the inter-war period,
mainly because neither the landlords nor the state were able or willing to offer relief
from rents and taxes (Scott 1976).

Tenancy is a commercial concern when the landowner has a large property with
many tenants and an agent may handle the business. In late eighteenth-century
England, both long-term leases and annual rentals of farm land were commonly
handled by lawyers, although the landowners' stewards would inspect the land to

ensure that contract terms had been met, and to determine the rent for the next tenant, or the next lease period for the same tenant (Middleton 1798). Where land is abundant in relation to the number of clients, rents are usually low. But where there are many clients and little land, half or more of the produce of the farm, or its money equivalent, may be demanded as the price.

Often, tenancy involves a personal relationship, friendly or otherwise. Many landlord–tenant relationships are essentially of patron–client type, in which there are stated or unstated obligations on both sides. These obligations are, or were, the essence of what Scott (1976) classically described as the 'moral economy' of peasant society. On the landlord's part, they involve loans to the tenant at times of need, willingness to defer or reduce the rental payments of tenants when crops are poor or fail, and support with and protection from the demands of the state authorities at all times. The tenant, in return, does free work for the landlord, makes small contributions to social payments which the latter is obliged to make to sustain his community or kin role, and gives him physical support in disputes.

Tenancy and the product of labour

All forms of farm tenancy involve, in effect, an exchange of the product of labour for land. A tenant paying a fixed rent, in cash or kind, is providing the landlord with a part of the product of his labour in return for access to land. For a sharecropper, the labour element is more specific since income earned off the farm cannot be substituted for the crop share. For this reason, we discuss sharecropping in this chapter in terms of the land transaction only, and in relation to theory. The labour aspect of sharecropping is reviewed in Chapter 4.

Tenancy arrangements are often discussed from the point of view of their economic efficiency, with the underlying belief that security is the most important condition for innovation, investment and progress. Fixed rents, whether in cash or kind, permit both landlord and tenant to know where they stand, and this is thought to be good for business-like decision making. They are certainly good for the landowner, but for the tenant there are disadvantages with fixed rents, even if the contract is a secure one. Rents paid as crop deliveries were normally paid after the harvest, as with sharecropping.[4] Some landlords seek a year's rent (or more) in advance when it is payable in cash. Under stable economic and social conditions, the granting of long leases to tenant farmers, leases covering from 21 to 99 years, became a common practice in Europe, and long and renewable leases were not unusual in Asian countries. Rapid economic, social and technical change lead to severe disruption. In Malaysia in the 1960s and 1970s, leases for periods of three or more years came to be paid wholly in advance and were therefore made only by well-established farmers or entrepreneurs who could afford these lump sums. In a period of rapid commercialization and mechanization, small-scale tenants rapidly shrank in number (Scott 1985).

Sharecropping remains widespread in West Africa, where the normal rent share is either one-half or one-third of the principal crop. There is a variety of arrangements, and in some of them two-thirds of the main crop goes to the landlord, and less or none of minor crops. Sharecropping is most commonly applied in areas where there is substantial immigration from other regions, but tenancy arrangements apply also to locally-born farmers wishing to enlarge their holdings or work land in closer proximity. In one southern Ghanaian community in 1999, farmers farmed freely on land inherited from their forebears, but 23 per cent rented additional land, while 62 per cent also hosted tenants on their land (Gyasi *et al.* 2003). In the Brong-Ahafo region of central Ghana, new migrants from the north are introduced to landowners by resident relatives, and then get land for one-half or two-thirds of the crop, together with a traditional payment to the chief which applies to all farmers in the locality. Others are 'adopted' by patrons in return for farm labour and other services rendered to the landowner, but pay no rent beyond the payment to the chiefs (Sarfo-Mensah and Oduro 2003).

Farming is subject to a range of indeterminacies, not least the serious risk of partial or complete crop failure. It is possible to reduce the indeterminacies to a statistic over the long run and in retrospect, but there is no possibility of being certain about the outcome from one season to the next. While coping strategies include diversification both of crops and of enterprises, the farm rent is usually based on the principal saleable product of the land, or its expected value. Other things being equal, the tenant will usually prefer an arrangement under which the risk is at least shared by the landlord, and not wholly borne by the tenant farmer. Sharecropping would seem to provide this flexibility. Scott (1976) argued that the southeast Asian tenant farmer, and by extension all tenant farmers, will evaluate tenancy arrangements by the degree to which they provide, or fail to provide, insurance against the risk of crop loss, price decline or unexpected increase in the cost of inputs. A variable rent, as in sharecropping, would seem to provide better insurance than a fixed rent. Why then has sharecropping been so detested by so many writing about agriculture?

The debate over sharecropping

Ever since the time of Adam Smith and Arthur Young, sharecropping has been a particular target for the wrath of economists and agricultural writers. Neo-classical economists have argued that as the sharecropper receives only a fraction of the return for their marginal inputs of labour, there is a strong disincentive for effort. The arrangement diminishes the incentive for landowners to invest in improvements. Marxists see it as extraction of surplus from the tenant, and as an obstacle to the evolution of capitalist farming with wage labour. Modern economists likewise see sharecropping as constituting a brake to the concentration of land in the most efficient hands. In the USA, sharecropping is particularly associated with the devices employed by former plantation owners to obtain continued production from their former slaves after emancipation in the 1860s. By the 1880s, big areas were worked by mainly black sharecroppers, often bound by debts to the landowners. The system extended also to farmers who were not former slaves and to other areas, including the Midwest.

Robertson (1980), an anthropologist, discussed the objections to classic sharecropping and noted that there are advantages for both parties to any sharecropping arrangement. Usually, the tenant gets land which he could not otherwise afford, and the landowner can get income from land which he lacks the resources to work himself. For the small tenant there is the obvious advantage in sharecropping that the rent diminishes when the crop is poor, something achieved under fixed rents only by the bounty of the landlord. Viewed internationally, there are wide variations in the nature of the generally informal contracts that are involved, demonstrating a flexibility that characterizes all such interpersonal arrangements. As we will see in Chapter 4, modern sharecropping has taken on very different and advantageous forms.

Reverse tenancy

An old practice has grown in importance recently, under which the tenant is the big farmer and the landlord the smallholder. By this arrangement large and growing farm enterprises acquire additional land, while the smallholder is able to reduce his farming, or exit it entirely, while still retaining rental income from the land. A good deal of modern enlargement of farms has been achieved in this way. For reformers, in the European Union for example, who wish to see farms grow larger without small farmers being driven to the walls of bankruptcy or repossession by mortgagors, reverse tenancy has attractions since the departed farmer can retain rental income to add to that from new employment, or a pension. Quite often, on the other hand, the landlord becomes a wage employee on his own land and the combined income remains small. Data on the amount of reverse tenancy are sparse in the extreme, but the practice is world-wide, and we will encounter its consequences in later chapters.

The complexities of landholding

From borrowing to unimpeded full ownership there is a continuum of practices, not easily subdivided. It includes the many different forms of sharecropping, commodity rents, cash rents and leases of short and long duration. It does not seem profitable to draw lines within this continuum. It is reasonable to separate owned land from rented or borrowed land, but there is a great range of freedoms and restrictions within any arrangement. Sharecropping, and rent paid in kind, shade into other forms of rental and labour supply so that it is of limited value to distinguish sharecropping as a separate category. Even where there is individual ownership, it may be hedged around with certain prohibitions and responsibilities. There may not be freedom to alienate the land to persons who are not members of a particular group, and there may be restrictions on the form of land use that can be practised. Always, actions that affect the land of neighbours are either subject to their prior agreement, or may be the cause of costly dispute.

Nineteenth-century liberals, and some modern neo-liberals too, thought of landholding as the simple control over all rights, with complete freedom to dispose of land at will. The fact, as has been recognized in modern times and as our discussion demonstrates, is that rights in land are a bundle within which only some belong without restraint to the holder while others are subject to restriction by custom, contract or law. This applies across the whole range of systems reviewed in this book, and has held true throughout history. The late twentieth century and the present decade have seen further sharpening of these restrictions due to state and international policies. The special case of water rights is discussed below, and in chapters 9 and 10 we will see how modern notions of environmental stewardship have strengthened the hand of public rights as against those of the individual.

Access to common land

Some resources are held in common, even in the most privatized of systems. Commons have individual users but are not subject to individual possession. Common land has diminished greatly, but in the past it was of great importance for the livestock of the poor, and of value for pasturing livestock during the grass growing season by adjacent farmers having common-land rights. Even in Europe, large areas of upland pasture are still held in common, usually by a defined group of farmers. Common land also serves as a source of fuelwood, constructional timber, leaves and grass for fodder, bedding or compost, wild vegetables and fruits, and for hunting and trapping of wildlife. There has been pressure to privatize common land in the now developed countries for more than two centuries, and once this happened it ceased to be available for these free-access purposes. In modern Bangladesh, the landless very poor subsist not only by selling their labour to the landed, but also by collecting uncultivated food from common land and from the sides of fields, which also yield them fuel and feed for their chickens. Enclosure of common spaces threatens their very basic livelihood (Frida Akhter, unpublished, quoted in Brookfield and Padoch 2007).

There is a huge literature about common property rights, much influenced by a simplistic but very influential paper by Hardin (1968). Individuals are assumed to have no incentive to 'stint' their use of the commons, since the degradation produced by an additional animal is borne by all, but the benefit is gained by the individual. Hardin considered this to be normal practice on common land, and thus we have the 'tragedy of the commons'. As Hardin and those who have followed him see it, the solution is either command-and-control direction, or else full privatization. A subsequent literature emphasizes the difference between 'open-access' commons and common resources with restricted access, which can be managed by regulation or collective action. Many writers have described the ways in which common rights are managed by those with access to avoid the 'tragedy' (McKay and Acheson 1987; Ostrom 1990). Unfortunately, not much of this writing pays attention to the vital importance of the commons to the landless. Meanwhile, the drive to privatize common land has continued.

Narpat Jodha (1987) provided a detailed account of what happened to common-property land in semi-arid Rajasthan, northwest India, following a land reform in the 1950s. Previously, landlords had regulated access to the common pastures through a system of local taxation. This system was extinguished, and large areas were enclosed by individuals practising arable farming with livestock; control over the remaining commons was left to village councils. Over the subsequent years to the 1980s there was major loss of productivity from the common land due to serious degradation. Those without bases of individually owned land have been much less able to profit from new commercial opportunities, and many have been reduced to tending the livestock of the well-off as wage workers.

Rights and property in water

On the face of it, water should always be thought of as a common property resource. An insightful jurist, cited by Anderson and Hill (1975: 176), ruled that 'For water is a moving, wandering thing, and must of necessity continually be common by the law of nature, so that I can have only a temporary, transient, usufructary property therein.' Yet, transient or not, rights in water are keenly argued as property rights in many parts of the world. On land adequately fed by rain, water is a free gift of nature, although this does not prevent farmers from seeking public or other assistance in time of drought. In drier areas, land is of little value without the artificial provision of water, whether by diversion of natural flow or by abstraction of groundwater. In these regions, water rights rather than land rights become critical. Control over water can mean control over land, as in the western regions of Botswana where access to modern boreholes has given the privileged few an oligopoly over large areas of semi-arid pasture (Rohde *et al.* 2006). Water rights are also critical where reliable additional quantities of water are needed, particularly for the cultivation of wet rice. Access to water is far from the only problem that has to be resolved. Land has to be protected from floods, the sea and invasions of salt water. It also has to be drained. Of China, Elvin (2004: 115) wrote that 'water-control systems are where society and economy meet the environment in a relationship that is more often than not adversarial', although he might also have been referring to the Netherlands and a number of other regions. In most cases, however, water is a socially managed resource.

Managing water

The social management of water is enormously affected by considerations of scale, and of the dynamism of the natural processes being managed. Only on a small scale can water be managed by a single farmer, without reference to others.[5] A whole community is usually involved, while larger-scale management brings in government at levels from the local to the state. By a commonly cited World Bank estimate, however, 85 per cent of irrigation world-wide is managed by associations of the water-users, with command areas ranging between one and many thousand hectares (Mabry 1996). Some of these associations constitute an effective local government because their officials, necessarily involved in adjudicating disputes concerning a critical resource, also adjudicate in other matters. Other associations can be invisible to all except those involved.

Until the 1970s, the 'water temples' of Bali in Indonesia were such an invisible system. A complex and very beautiful system of rice growing terraces is managed by farmers' organizations termed *subak*, of which there are 172 along two small rivers alone. Bali is a volcanic island, subject to a strongly seasonal climate, and with episodic major droughts due to the El Niño variation. High rainfall in the northern mountains feeds a large number of southward-flowing rivers which are also fed from large groundwater flows in the deep lavas. Over a period of about a thousand years, these rivers have been tapped by weirs feeding diversion tunnels and open channels which in turn feed the *subaks*. Before the 1960s, these permitted two crops a year separated by a short fallow period. The Balinese form of Hinduism recognizes a water goddess and an associated ritual is practised at shrines above each *subak* and above each weir. Embedded in the ritual are mechanisms determining the allocation of water and the synchronization of cropping patterns. The demand for water peaks at the beginning of each cropping season when land must be puddled and then flooded to receive the transplanted rice seedlings.[6] To ensure that this could be done without moisture stress, different *subaks* or groups of *subaks* staggered the beginning of each cropping season by up to two weeks, and if there was insufficient water to meet the demand, all *subaks* in a command area received a diminished allocation, agreed among the temple congregations and their representatives at the command area temple. Other responsibilities of the congregations included adherence to the agreed cropping system, and work on maintenance of the whole system.

One thousand years of this system were terminated in the late 1960s by the Indonesian Green Revolution and its authoritarian 'mass guidance' programme. With high yielding rice varieties and fertilizer it was determined to produce three crops a year, without reference to the old system of coordination among *subaks*. By the mid-1970s the system was in chaos with great yield variations and also a major attack by insect pests, flourishing among the almost unbroken supply of rice plants. Farmers, who had sustained the religious rituals, began in the 1980s an attempt to recover enough of the old system to permit sufficient synchronized fallow to break the pest life cycle. Their 'superstitious' ways were decried by the authorities but an American anthropologist, Steve Lansing (1987, 1991), was able to persuade the latter of the values of the old system. By this time the political regime that had embraced the Green Revolution was starting to break down, and some farmers were able to return to growing only two rice crops a year while others continued with three. At this stage, Lansing and Kremer (1993) modelled the consequences of water-temple decision making in two catchments, incorporating pest reduction as well as regionally optimal water allocation. They were able to show that a stable solution might be reached through local coordination and adaptation within only a few years. Further work (Lansing and Miller 2005; Janssen 2007) has confirmed and elaborated the value of the water-temple system, and at the same time established that such a bottom-up system of irrigation management is superior to anything that central authority might have come up with in the past, or that it has yet produced with all the advantages of modern science and technology.

Computer simulation may have mimicked history, but it cannot replace it and not everyone has been prepared to agree with Lansing and his collaborators that such a complex system could have arisen without authoritarian command. In northwestern Spain, by contrast, documentary evidence goes back to the first re-occupation after conquest from the Moors in the early Middle Ages. In a valley in the rain shadow of the Cantabrian mountains, a complex system involved groups of villages in irrigation associations termed *presas*. The *presas* collaborated to share the substantial costs of construction and maintenance of the infrastructure, and the legal costs of sustaining collective rights, which have survived into modern times. In the Orbigo valley, Guillet (1997, 1998) has analysed the historical evidence in geomorphological, legal and social context. The Orbigo, like Bali, is an example of 'cascade' or 'chain' irrigation, in which the excess or return flow from one user or group of users becomes the intake for the next user downstream.[7]

Although unresolved disputes could go to law, most of the many disagreements are resolved locally. The system that evolved in medieval times entailed use of an old channel of the Orbigo to carry the main canal through the participating community lands, and downstream to users beyond. Collaboration included breaking down the dam on the river and partial blocking of the main intake each winter to protect the system and its villages from the early spring floods. The infrastructure had to be rebuilt at the start of the next irrigation season in the late spring in time for peak demand when land was being prepared for crops. Flows were unmeasured and continuous, permitting flexibility in water use according to changing needs. Each municipality had its own diversionary system from the main canal, under agreement which involved sharing of the costs and the work of the whole system. They then made water available to end-users on the basis of use-rights attached to each parcel of land, and not saleable separately from the land.[8] When water was ample, supply was subject only to queuing; when it was scarce a system of turns applied, sometimes supplemented by a lottery. Water mills, which each community needed, caused many problems so that most of them were erected on the exit canals from each sub-system where interference with irrigation was least. The mill owners could sell or rent water from their own channels for minor irrigation or other purposes. This went on until 1951 when a state-built dam on the upper river controlled the flood problem and also ensured a more secure supply in the dry summer months when rationing had been necessary.

Under a national water law of 1879 *presas* could apply to become *comunidades de regantes* (irrigation communities), giving them firmer legal rights, especially to the rights of way for water that had often caused dispute with upstream neighbours in the past. By 1975, all *presas* on the Orbigo had become *comunidades*, 18 in number, irrigating some 10,000 ha. Subsequent attempts by the national authorities to enforce measuring with a view to making charges for water have been frustrated not only by opposition, but also by the high cost of installing devices that are easily sabotaged and cannot measure actual net consumption without major system reconstruction. Although water is now released from the dam on a system of turns for the *comunidades*, the system continues as before on the basis of continuous flow and flexible access, with prior users supplying their excess to downstream users. These latter, who include irrigators outside the *comunidad* system, have benefited from canal lining and other improvements that augment flow. Arguing against demand management and market allocation of water more generally, Guillet (1997,

2006) uses the Orbigo and a Peruvian case he has studied to make a case for the efficiency of the many hundreds of thousands of farmer-managed systems, meeting irrigation and other demands which, like natural supply, pulsate strongly and often unpredictably through the year. Moreover, these farmer-managed systems remain decentralized. 'Adjudicative or central authorities, and hierarchical or bureaucratic organization are not necessary for the system to work'[9] (Guillet 2006: 327).

BOX 3.2 WATER ALLOCATION AND SOCIETY IN A SRI LANKA VILLAGE

Leach (1961) has provided perhaps the most detailed account anywhere of the tenure of land and water in a single village. He writes of a Sri Lanka village in and before 1954, before the time of the Green Revolution. Pul Eliya received two periods of rain each year, the second more reliable than the first. Its irrigation was independent, forming no part of the 'chain' or 'cascade' systems of central Sri Lanka that were centrally managed in the ancient past and are centrally managed today. It had one main reservoir (tank) and, before colonial period addition of some freehold and leasehold blocks, one main field fed from the tank. This field was divided into three roughly equal sections, two in the upper field which had the most reliable water, and one comprising the lower field which was entitled to only one-quarter of the water. Each part-field was further divided into strips at right angles to the channels. The strips were the holdings of shareholders, each with a separate entry from the channels. The shareholders' water rights were so distributed that those with best access to water also had land that was less well supplied. The land rights were water rights, distributed on egalitarian principles in the old field, but not when the newer additions were taken into account. Control over water was exercised by a notionally elected water officer, the *vel vidane*, who would open the entry channel from the tank on a predetermined date. Water division was achieved by use of grooved logs in the base of the channels, the grooves being of precise but unequal width in order to allocate the flow of water according to anciently determined proportions.

Land preparation began in advance of the opening of the sluice, but once there was water in the channels had to proceed swiftly through ploughing, harrowing and levelling the plots. The plots were internally divided into rectangular basins of which the shareholder could use as many as he wished, or as there was water to supply. Rice seed was broadcast into these divisions, with no transplanting being involved. Water levels were maintained through the growing season at a height sufficient to keep the rice ears above water. Water supply ceased when the grains ripened, and harvesting was undertaken by both men and women. The subsequent threshing, with buffaloes, was done by men only in cleaned patches of the field, most of them conveniently located to carry the threshed grain to the village.

This discussion of the place of water management in farming is far from exhaustive. A simpler, smaller case is developed in Box 3.2. Water is, however, an element that has to be applied to land, just as labour, tools and livestock are a mobile property of the farm. We move on to these elements in Chapter 4.

Notes

1 The '*formen*' is a shorthand for '*Formen die der kapitaliistischen production vorhergehen*'.
2 The inferior position of women in pre-communist China has already been touched on in Chapter 2. Female descendants had no claim to property and, although a widow would often control land after the death of her husband while there was no adult son, it would pass only to male heirs. Though legislation changed this in the 1930s, it was not effective, and only after enactment of the marriage law of 1950 did women inherit equally with men (Yang 1959a: 141). Within a few years these rights, so far as land was concerned, were extinguished by collectivization.
3 Brookfield worked in the Chimbu area of Papua New Guinea in the 1950s and 1960s. At that time, several blocks of arable land in the central part of a 2 sq. km subclan territory were, in the absence of any cadastre, attributed to a small number of men who were leaders in every enterprise. Much of the land was worked by younger men, either members of the same group or married to women from the group. To all intents and purposes the 'owners' were the landlords of the younger men they sponsored. In the 1980s, this same land had become attributed to the younger families, so that in effect they have become the owners. No rent or crop-share had been paid, but the now elderly former 'landlords' continued to enjoy both respect and support (Brown *et al.* 1990).
4 Tithes, where they continued to be paid, remained crop shares (normally one-tenth), and in England at the end of the eighteenth century most clergy insisted on taking them as crops, so that they could resell them for their own profit (Middleton 1798).
5 There are probably half a million farm dams in Australia, mostly for the watering of livestock. They intercept water that would otherwise flow into the rivers, and in time of water shortage they attract a significant degree of criticism. Nonetheless, the right to impound water on a farm is rarely disputed, in Australia or anywhere else.
6 In Japan, at least until recently, group work persisted in the planting out of rice seedlings because this had to be done quickly. Since irrigation water could not be supplied to all fields at the same time, groups had to work from field to field as the necessary flooding water was supplied to them, for a matter of hours rather than days (T.C. Smith 1959: 53).
7 Water is lost by evaporation, domestic or animal consumption, seepage from unlined canals, and evapotranspiration from the cultivated plants. The rest flows on until returned to the source river or to a canal taking it to the next user. Necessarily it will be charged with salts leached from the ground and with pollutants added by the prior user. In modern times, pollutants have become a serious problem (Chapter 9). Except in times of low flow, normally at least half the first intake is passed on for subsequent use.
8 The uncomplicated device of attaching water rights to land is widely used in the world's irrigation regions. It means a right to a share in available water rather than to a specific quantity, although quantities have been specified, at least annually, in some large-scale modern administered systems. It has been bitterly attacked by those who believe that the proper way in which water should be allocated is through the 'free' market, so that it can find its highest value. In Australia, this debate has become very political in the face of a long and severe drought in the years since 2000. Urban claims are advanced as superior to rural claims, and irrigation farmers bitterly dispute the allocations to upstream farmers, especially when these competitors are cultivating

crops with high water demand, such as cotton in the northern reaches of the Murray–Darling system. Whether for this reason or because of the general severity of drought, downstream users, both urban and rural, can find themselves bereft of supplies.

9 Guillet relies on Schultz's (1964) 'poor but efficient' farmers as his model, but Sivakumar's (2001) model of satisficing farmers, discussed in Chapter 1, is perhaps more appropriate.

4   Workforce, livestock, tools and seeds

The family as workforce

By definition, the core workforce on a family farm is the family. From a traditional view, when the family is young and time is taken up for some years with raising children, the male farmer is the only regular workforce for the farm. Later, his wife is able to take a larger share in work in the fields and with livestock, and sons and daughters also participate. In Chayanov's scheme of things, discussed in Chapter 1, this is the period when the family workforce is rising toward its maximum, and the producer/consumer ratio is maximized when the sons are fully adult, but have not yet set up households of their own. Chayanov to some degree, and writers who have followed him more so, assume that family labour is unpaid, but there are many forms of payment other than a regular wage, and we do not treat either family labour or the help of neighbours as without cost.

Farmer and family

Until quite recent years, when male off-farm work has become more common, the 'farmer' in the literature has almost always been presumed to be a man, generally the 'head of household' who is assumed to be the sole decision maker. Even in official statistical collection, he has often been assumed to be the only worker on the farm other than additional adult males hired for short or long periods. Historical data runs concerning female and adolescent employment on farms have yielded little in the way of true information to modern social scientists and historians. Yet women and even young children have always been involved in work on the farm as well as in the home. The only systematic exception has been in countries and regions where adult women are socially segregated from men, or excluded by custom from work in the fields. Widely there are ascribed gender roles, in which women take responsibility for specific areas of activity outside the household, such as care of the livestock, while field labour is ascribed to males. In others, women cultivate the food crops, men the cash crops. In north-central Sri Lanka, as we saw in Box 3.2, men did most of the work, but women were specifically excluded only from the threshing. Societies like the Kofyar of Nigeria, described by Netting (1968) and Stone (1997), in which almost no tasks are gender specific, are comparatively rare. Moving into a more spacious environment in which commercial production of yams became a dominant activity, the Kofyar quickly adopted organization into larger households. Such enlargement of the household in the face of a relative shortage of labour reflects an ancient practice.[1]

Whether the household is large or small, and whether or not there are socially sanctioned male and female tasks, individuals have to know what to do. How does this happen? The family table, or bowl, round which final decisions on the next day's work are most probably made, or at least confirmed and ratified, remains a 'black box' in the family farm literature. What is clear is that the old implied notion that all decisions are taken by the patriarch, and simply handed down and received without demur, is only sometimes the reality. Except perhaps in China, where the family has been more focal to society than in any other world region, patriarchal control has probably only been the reality where strong sanctions could be imposed on junior members of the family. An appointed farm manager has to do something very like this, but families are inherently less autocratic.

Whether as equals, or in a more subtle role, women take a large share in making farm decisions as Netting (1993) stressed. Brookfield recalls a lunchtime discussion at the farm on which he worked as a seasonal labourer (Chapter 2). The male farmer declined to take his wife's advice, but then followed it in conducting and directing the afternoon's work! Nonetheless, the common practice in most countries of treating the male as head of both household and enterprise has had its effect. One consequence of mechanization has been to enhance the significance of male labour, and of male authority (Saugeres 2002). In the USA and Germany, Adams (1988) and Prugl (2004) recount somewhat similar modern stories. In the early twentieth century, women continued semi-independent activities with small livestock, vegetables and fruit, and the processing of milk products. They marketed the product themselves. Their contribution to field labour was principally at peak times, as 'working farm wives', but not as fully acknowledged members of the labour force. This partial independence was eroded during the 1930–60 period by competition from specialized farms, and by a drift toward the middle-class urban model of non-working wives, which was only after 1960 overtaken by the feminist movement that sought greater equality for women, including paid employment by choice and not only from necessity. Prugl (2004) adopts the feminist term 'women farmers' to describe all women working in farming as opposed to only housework, irrespective of the scale of their input. In fact, in Germany as also in Australia (Garnaut *et al.* 1999), modern more gender-sensitive inquiries have shown that the total working week of women farmers, including housework, is commonly greater than for men farmers, and often exceeds that of urban women who have jobs as well as main responsibility for the household work. Women often handle the farm bookkeeping, and very commonly also all or most direct marketing, but it is futile to try to distinguish what is a central contribution from what is a purely subordinate one.

The labour of children is less discussed than that of women, and the data are even less satisfactory. Teenage children are normally employed either in farm work or in household work, and the supervision of livestock, the collection of water and firewood is often carried out by children. Even young children are commonly employed in the fields at periods of peak labour demand. The global trend for education to extend over longer periods has both reduced the availability of teenage labour and also provided teenagers with the means to find and obtain off-farm jobs. Modern education systems are designed by urban dwellers, largely for urban dwellers, and teach little that is relevant to rural needs. Education is the least questionable element of the 'urban bias' in rural development of which Lipton (1977) complained, yet one reaches page 257 in his controversial book before coming to a discussion of the role of education in driving the rural skill drain. In modern work ethics child labour is a 'bad thing', and its prevalence in farming is either the subject of critical comment, or is tacitly ignored.[2]

For the present, we have to leave the organization of the family workforce not much beyond the 'black box'. Ellis (1988a) is almost unique in applying the reasoning of modern household economics to the specific question of the choices facing women on farms. He concludes that 'social custom and obligation predominate over individual choice and decisions do not correspond to marginal utility criteria' (Ellis 1988a: 181). Only doctrinaire economists who think all economic choices to be individual would believe otherwise.

Hiring in: the seasonality problem

Only the least complex of family farms can carry out all its operations with family labour alone. Because agricultural work is highly seasonal in almost all environments, and because both the natural and social conditions of production and marketing vary substantially through time in both predictable and unpredictable ways, flexibility in farm use of labour is essential. Sufficient flexibility can rarely be provided within the farm family alone. The simplest way of coping with this problem is by exchange of labour between farms on a reciprocal basis. Large jobs that have to be done in a short time period, such as harvesting, ploughing, seeding, transplanting, clearing a long fallow or a new field and maintenance of infrastructure, can better be performed if neighbours and kin assemble labour forces on one farm at a time. As in the Japanese and Nigerian cases described in Chapter 2, there is no immediate payment other than provision of at least refreshment for the workers by the hosting household; payment comes when the contribution is reciprocated, or reimbursed in kind. Reciprocity still happens, and where full reciprocity is not feasible because of the different labour endowment of farms at different stages in the family cycle, arrangements akin to sharecropping can supplement the free loan of labour time. Labour is also provided on a less voluntary basis to households that stand higher in the social hierarchy, in return for patronage at a later (or earlier) time.

Such arrangements do not help where longer-term support is needed, or as farms have grown larger. Hiring of labour for wages for short or longer terms became the practice long ago in the now developed countries. Wage labour (and slave labour too) had been used on farms since classical times, but it had always been associated more with the larger estates than with smaller farms. In Europe and North America, a substantial class of permanent wage farm workers did grow up in the eighteenth and nineteenth centuries and persisted well into the twentieth century, but its members were then the first to take opportunities to move into other kinds of work. Permanent wage workers still exist on family farms in some developed countries, but whereas larger family farms might employ only one or two, it is only the industrial farms that depend mainly on their labour. One would not nowadays find two regular labourers on a 40 ha general farm, such as the 1940s Gloucestershire farm described in Chapter 2.

Since rural labour requirement is seasonal in most climates, maintenance of an employed labour force adequate for the main cultivation season would necessarily involve supporting people who would be partly idle during the off-season, if they had no other employment. Permanent employment of unskilled labour could only be economic to farmers who had a year-round mix of activities. Modern trends toward specialization, together with mechanization, wiped out the basis for year-round employment, although the consequence is often part-time employment of costly machines. For smaller farmers in many countries, seasonal labour could and still can be obtained from among the local non-farm community, or from among the smallest and poorest farmers. Sometimes this is done on a regular basis, the same workers taking seasonal jobs on the same farm in successive years. Such arrangements have been in steep decline wherever continuing non-farm employment has become widely available. Other forms of obtaining seasonal labour have evolved, and are discussed below.

Specialists and sharecroppers to whole-farm contractors

Employment of skilled and specialized workers has always been expensive, and for the smaller farmer their services were needed only intermittently. Therefore, they were engaged mainly on contract, providing a service more than labour. The farmer who hires out his working livestock or machinery, and himself, to other farmers both obtains income and provides a service. The critical task of sheep shearing on most Australian wool farms has been performed for a long time by contracted groups of shearers whose visits are major events in the annual cycle. Formerly, they received full board in shearers' quarters on large properties, in addition to the cash wage. Nowadays, they stay more usually in town and commute out to the farms to work. On arable farms, seasonal harvesting labour has been almost completely displaced by the harvesting contractors, with or without their own workforce. Generally, the contracting of specific tasks has become increasingly common as specialized equipment is held by specialists, while the widely capable general labour of former times is decreasingly available.

We saw in Chapter 3 that sharecropping, as an institution, has remarkable persistence. It has changed in nature as the decline in social inequality has reduced that aspect of the relationship which made the share-tenant a de-facto labourer for the landlord. At Kossho in Japan around 1910 (Chapter 2), Nagatsuka wrote of indebted tenant farmers that:

They worked for long hours in their fields doing all they could to raise enough food. Then after the harvest they had to part with most of what they had produced. Their crops were theirs only for so long as they stood rooted in the soil. Once the farmers had paid the rents they owed they were lucky to have enough left over to sustain them through the winter.

(Nagatsuka 1989: 47)

While sharecropping still flourishes in many developed countries, it has sometimes evolved into a very different form. In regard to the American Midwest, its theoretical advantages and pitfalls have been keenly reviewed by several writers, notably by Reid (1977), who stresses the advantages. The contract joins landlord and tenant interests, and on an entirely normative set of assumptions is advantageous to both. Sharecroppers renting land on almost a thousand farms in Illinois each have a full suite of large farm machinery and contracts with several farmers, so that the area worked by one operator may be much larger than one farm (Young and Burke 2000). The share arrangements, nonetheless, are much as in the past and in other parts of the world. The 'tenant' and 'landlord' each get either half of all crops, or in areas where the expected returns are lower, two-thirds and one-third respectively. These 'tenants' are experienced and well informed; they negotiate on an equal basis with the 'landlords'. Although there is some variation in the proportion of fertilizer and other inputs provided by each party, there is only limited departure from the crop shares that are customary. Sharecropping in this form also flourishes in Australia where older forms of sharecropping have persisted. Under the modern arrangements, the 'tenant' and 'landlord' both gain: the former farms without owning the land, and the latter gains a product (e.g. grain or hay) without owning machinery or investing labour.

Commonly, the landowner does not use the rental crop for household consumption, but sells it. In a Catalan village in Spain, before the 1950s, 133 tenants rented land from 57 landowners, paying half of their wheat and most other crops (Asano-Tamanoi 1988). This was a direct and personal relationship between landlord and tenant, and the tenant's benefits included wage employment in the landlord's other enterprises in the off-season. Landlords sold the rental crop, thus paving the way for the contracts with European multinationals which began to seek deliveries of feed crops and meat in the late 1950s, accelerating a change in land use and in tenurial arrangements. There remained 57 landowners in the Catalan village, as in the 1930s, but numbers of tenant farmers declined and they now pay rent in cash. In the 1970s the rental sum was based on the price of feed cereals, by that time the principal marketed product. The partnership in production and marketing formerly managed through the tenant/landowner relationship instead came to be managed between the farmer and the company agent on a contract basis.

There is also a profession of whole-farm contractors who take on total management of the farm during long-term absence, illness or semi-retirement of the owner. Other contractors specialize in tasks such as construction of modern livestock facilities, or fencing, although at a cost that deters many smaller farmers from using their services. The distinction between sharecropping and contracting is as full of nuances as any other discussed in these chapters. In the developed countries, an enterprise may begin by contracting, say, the harvest, using a rented harvester. The entrepreneur may then move to whole-farm contracting or sharecropping, then renting, and later acquire his own land. It is a new first step on the 'agricultural ladder'. Contracting and sharecropping are related in various ways to marketing. A whole crop may be sold before harvest to a contractor who then hires his own labour, and uses his own machinery, to harvest it. This system arises not only in developed countries, but is an old-established practice with the rice crop in countries such as Indonesia. Not dissimilarly, in New Zealand 'share-milkers' milk the farmers' cattle and sell a share of the product; there is a variety of arrangements (Smith and Montgomery 2003). None of these arrangements is 'archaic', as sometimes described; they operate well to get jobs done in an efficient manner at reasonable cost, and contain few elements of exploitation. It is a rather different matter with arrangements that would be described by some as both modern and 'economically efficient'.

Migrant workers and contract gangs: the modern proletariat

Unskilled and semi-skilled farm workers have durably been sought from two sources: smallholding farmers whose own land is inadequate to provide full family employment with sufficient reward and seasonal workers who move from areas of limited opportunity to areas with greater opportunity. There is a significant overlap between the two groups, in that all have insufficient land or local income, whereas the migrants among them also have insufficient nearby opportunity. Thus Welsh and later Irish seasonal workers were hired in southeastern England in the eighteenth century, and even earlier. Breton and Spanish migrants worked in summer in the south of France. The 1940s *bracero* migration of Mexican workers to the USA was the precursor of present-day dependence of much American farming on labour migration, legal or illegal, from many parts of Latin America. The seasonal migration of Polish workers to farms in Germany and France was a major element in the period before 1940. Since the 1990s, several thousand Polish farm workers have been welcomed in Britain.

In most of the twentieth century, until the 1960s in some developed countries and later in others, farmers hired migrant workers on an individual basis, contacts being through the migrant networks. In northern Spain it was still like this at the end of the century (Hoggart and Mendoza 1999). An alternative, still common in the intensive farming areas of Mediterranean Europe, is for migrant workers to gather in a town square, or other known meeting ground, at daybreak to be hired for the day by farmers. Larger farmers, and tropical plantations, prefer to hire their labour through agents, and to hold them for longer periods. The employment by larger farmers of mainly migrant agricultural labour has a long history, and it is largely a history of exploitation of the poor by the better-off. While some of the worst exploitation of which there is record seems to have taken place in the inter-tropical Americas, the British Agricultural Gangs Act of 1867 controlled some almost equally iniquitous practices among British 'gangmasters' who hired children, young people and women for agricultural labour. In the UK, women are still hired for casual labour in large numbers, and since 1970 this has been as contracted gangs by commercial farmers (Frances 2003). Since 1990, this locally recruited labour, sometimes hired from a considerable distance, has increasingly been replaced in Britain by foreign workers who are prepared to work more diligently and for longer hours to meet the demands for quality produce, delivered on time, to the supermarkets (Rogaly 2006).

The gangmaster system returned to Europe and North America in the later twentieth century in the guise of 'farm labour contractors', who arrange transport for the workers, pay them, and sometimes house and feed them.[3] Commonly they charge a fee of about 30 per cent of the wage bill, usually deducted from the gang workers' pay to recompense the gangmaster for finding them jobs and transporting them. A major reason for the attractiveness of this system to farmer employers is that it frees them from a growing host of labour legislation; the gangmasters assume this responsibility, and increasingly are required to be licensed, passing tests of competence. In jurisdictions unfriendly to farm worker interests it is possible for gangmasters to be quite lax in meeting these requirements. Until lately, most gangmaster operations have been small, and some of family-business scale, but in a highly competitive field there is great scope for corruption and favouritism, as well as for some iniquitous and illegal practices (Wood and Skaggs n.d.). These include links with entrepreneurs in the shadowy field of illegal immigration. Gang contractors in the USA have dealt principally with foreign immigrants from Latin America. Some British gangmasters also acquired an unsavoury reputation for their dealings with illegal Asian immigrants, which led in 2004 to greatly improved oversight of their activities. They deal also with European immigrants who are legal, though often not fully aware of their rights (UK Parliament 2003). Mainly supplying labour to the horticultural and fruit sectors, there are now some very large gangmaster companies in Britain (Rogaly 2006).

In the Western Cape of South Africa, rising costs and the responsibilities of new post-apartheid social legislation led many farmers to dismiss large numbers of their former resident workers. Remaining on-farm live-in workers are employed mainly on specialized skilled tasks, while a large part of the labour intensive work such as pruning, thinning and fruit harvesting has shifted to non-resident workers, often former residents who have lost their jobs (Kritzinger *et al*. 2004). Contract workers, few of whom have long-term employment, now do most of the field work in the orchards and vineyards of the region. Entrepreneurial contractors negotiate terms for specific tasks with the farmers and pay the workers a wage. Although there are some core workers who continue for a long period with one gangmaster, most are hired only for short periods with no security. For the fruit growing farmers, the system not only cuts costs but also permits flexibility in labour hire according to seasonal variations and market conditions. While it does not necessarily reduce incomes for the contract workers, it does create a high degree of insecurity and the risk of experiencing periods of want. In the world as a whole, only a small proportion of farm workers are unionized, and among gang workers trade union organizers find it very hard to make any headway. One other source of seasonal labour, quite outside the union system, is described in Box 4.1.

BOX 4.1 STUDENTS AND TRAVELLERS AS SEASONAL FARM WORKERS

One source of short-term rural labour with fewer problems than the gang system is students, hired during vacations. In recent years, with the great expansion of 'backpacking' among young people, this has become truly international, although with considerable discrimination – on, it would seem, ethnic grounds – among those seeking visas as 'seasonal agricultural workers'. Farmers' organizations, whose members are continually beset with seasonal labour problems, have lobbied for substantial enlargement of these flows, but encounter opposition because a proportion of workers on temporary work visas overstay and seek other work. In Europe and the Antipodes, at least, recruitment of tourist visitors outside these formal schemes is actively pursued, by advertisement through the backpacker hostels and by word of mouth. Without the same emphasis on students, seasonal migration continues actively in Latin America; flows from Bolivia to Argentina significantly augment farm labour supplies.

Notwithstanding hard conditions, opportunities for gang-like casual labour are eagerly sought in many parts of the world. In West Bengal, tubewell irrigation has now made possible two rice crops in a year. Since the 1970s, seasonal migration to the main rice areas in southern West Bengal has come to involve as many as half a million people at the peak of the monsoonal rice harvest, the busiest of four migration seasons in one principal district alone, Barddhaman, north of Kolkata (Calcutta). In one of several papers arising from work on livelihoods and migration in India, Rogaly and Rafique (2003) analysed one of four main migration streams to Barddhaman, drawn from near the Bangladesh border, where control over land and water, and hence wealth, are highly concentrated. The landless depend on paid employment in all seasons, but this is scarce and wages are low. Seasonal migration, of mainly Muslim men travelling without their wives, was for from ten days to about a month. It gave the migrants the opportunity to make lump sums, equal to from 10 to 25 US dollars, but in doing so both the men and their families endured a great deal. The men sometimes had to borrow the train fare for the 100-km journey. When they got to Barddhaman they had to negotiate with potential employers in a crowded informal labour market at the railway station since, unlike workers on some other migrant streams, they did not have jobs arranged in advance. Employment could be by the day, or by the task on piece rates which were popular with employers, less so with the workers. Returning with the riches of such employment, the landless migrant, and his wife, first had to repay the debts incurred to keep the family going during the workers' absence.

It seems very likely that the true rural proletariat of the twenty-first century will consist increasingly of the sort of people discussed above – seasonal migrants from poorer areas and countries, and members of contract teams or gangs created to replace former permanently employed farm workforces that have been reduced to save costs. The supply of part-time labour from nearby smallholders remains important, but has declined wherever other fields of employment remain open and under-supplied, so that farmers needing non-family workers are increasingly dependent on casual labour less experienced on the land and on gang specialists. The demand for casual labour is an inevitable feature of agriculture, and of commercial farming in particular. The gang system favours the larger growers who can provide the more secure business; in Europe and North America it has come to dominate the vegetable and fruit growing industries. In developed countries, the forces that have been strongest in the modern rise of the gang system of casual labour arise in the field of marketing, outstandingly the need for 'just-in-time' deliveries to processors, supermarkets and the fast-food industry (Chapter 5).

Livestock

The six main purposes for which livestock are kept on farms are draught, meat, their hides, wool or feathers, milk, eggs and manure. Draught animals have diminished in importance during the twentieth century as tractors have progressively taken their place, while the need for manure has diminished as chemical fertilizers have supplemented or taken its place. In the developed countries, except in areas of specialization, numbers of sheep have declined, especially sheep for wool. The demand for meat, milk and eggs has risen, and is increasing also in the Middle East and Asia. In developing countries, the new 'ladder' for many smallholder farmers is from eggs and chicken, through goats or sheep, to cattle. Cattle, and not only castrated male cattle, increasingly have been used for ploughing in some developing countries where tractors are too costly for most small farmers, or are unsuited to field conditions.

Most livestock breeds evolved before and during the nineteenth century, but there was still a great number of local breeds in the mid-twentieth century. They evolved, and many quite locally, by preferential selection and crossbreeding and through farmer interchange and sale. Today there is a much smaller number of major breeds and the genepool within these is contracting as well, with the high demand for characteristics that give potential for higher production and more efficient use of feed. The Holstein breed of dairy cows dominates throughout the world. In many developed countries, standardized systems of records generated through genetic and performance recording are now used as the basis of comparing and trading livestock. Identification of 'elite' individual animals and the use of the new reproductive technologies means a few individuals can come to dominate breeds. By the end of the twentieth century the price of a stud bull of preferred quality could approach six figures in euros or dollars. High cost excludes many smaller producers from the most productive animals. But mostly, livestock are selected and bought and sold on the basis of their appearance, through local markets or larger regional and seasonal markets, direct from known good sources or from neighbours, but also, in some cases, through schemes run by government or non-governmental organizations (NGOs) (e.g. the International Heifer Project). Price is determined by auction or otherwise by negotiation.

The complex chains from producer to consumer become more regulated when distant markets become the object of production. Wholesalers deal both in live animals and dressed meat, or the 'boneless beef' that the fast-food industry requires. Standards are imposed, whether at the international level of the Food and Agriculture Organization–World Health Organization (FAO–WHO) *Codex alimentarius*, or the higher standards demanded by some butchers, supermarket suppliers and other wholesale buyers. Disease control has always been a problem with livestock, and it has added greatly to the risks, as well as the costs, of going into livestock production. Infectious rinderpest and the tick-borne East Coast Fever wiped out entire herds of cattle in Africa at the end of the nineteenth century, and prompted the start of government regulation of livestock keeping practices in much of Africa in the early twentieth century (Beinart 1994). Foot-and-mouth disease was first reported in Europe early in the nineteenth century, leading to the alternative control practices of mass slaughter of infected herds, and of compulsory vaccination, in different countries. One major outbreak of foot-and-mouth disease which swept western Europe in the 1930s was initiated by a shipment of sheep from Morocco to France, whence it swept across porous borders as far as eastern Europe (Lamartine-Yates 1940).

The patchy state of disease control in Africa remains one principal reason why livestock ranching emerged in only a few African countries with well-controlled livestock systems, and has not been able to develop enough to compete with first the Antipodes and now also Brazil and Argentina in the flourishing meat markets of Europe, eastern North America and Japan. It leads to keen debate on how disease control is best managed (Scoones and Wolmer 2006). Control is linked to livestock movement and identification, and following new disease outbreaks in the late twentieth century the regulation of trade and movement has increased.[4] In Britain, Australia and several other countries, all cattle now have to have an electronic identification tag or lifetime bolus in the rumen. A modified system is being implemented for sheep. All these control problems add significantly to farmers' costs.

The tools

Before the limited introduction of steam power on developed-country farms in the mid-nineteenth century, almost all farmers had their own full suite of tools. Most of the massive steam-engine tractors, and other steam-powered machinery then introduced, were operated by an early generation of contractors. There were many successful mechanical innovations of family-farm scale. A large number of farmers, rural blacksmiths and mechanics invented new machines, and made adaptations and modification of inventions during the nineteenth and early twentieth centuries. Only a minority were commercialized, the most important being improved ploughs and seeding equipment, mowers, threshers and then thresher-binders and the first combine harvesters which emerged at the end of the nineteenth century. Farmers are still a source of new inventions. At least some inventions were made more than once, in different countries. From the early twentieth century on, a few successful manufacturers of farm equipment began to grow substantially, with purchase of smaller companies and then amalgamation of the larger ones. The greatest competition was among the makers of tractors which, by 1939, had already significantly displaced horse traction in the major regions of commercial arable farming. But the 1940s English farm described in Chapter 2 still worked with horses and only needed to hire in the services of a modern tractor-drawn and -powered threshing machine. Otherwise, this farm, like many others, still held its full stock of working tools.

After World War II, tractor power was applied to a much wider range of farm operations. Tractor numbers increased more than three times between 1940 and 1960 in the USA and, from a smaller base, by more than this in most of Europe. The USA and Britain in the early 1950s were the first countries in the world with adequate data collection to record more tractors than horses (Lamartine-Yates 1960). By reducing demand for manual labour while remaining affordable to smaller farmers, tractors increased the size of arable land that could be managed with a family workforce. Farm machinery dealerships, whether independent or tied to a particular company, had been set up in the country towns of all the main developed countries before 1939 and flourished greatly after 1945. Then, after the 1970s in a period of turbulent crop prices, concentration of ownership in the industry, accompanied by dispersal of much manufacture to low-cost countries, led to substantial reduction in the numbers of dealerships and in their freedom of operation (Pritchard 2006).

Since the 1950s, tractor horsepower has increased rapidly and the range of other machinery has continued to widen. The desirable range of arable farm equipment can now cost a sizeable proportion of total productive assets, and only large-scale farmers, families or not, can afford them without credit. Ownership and operation of the full range of equipment is increasingly in the hands of contractors in developed countries. On the other hand, there is growing penetration of mechanization in many developing countries, some of which, including India, China, Korea, Taiwan and Brazil, are developing and also exporting equipment of a more compact scale. The field remains highly competitive and there is also substantial trading in secondhand machinery and tools. The pages of rural newspapers are replete with advertisements for such equipment.

A note on seeds

The literature on farmers' seed systems has grown exponentially since the late 1980s, when the role of farms in conserving a diversity of germplasm was first fully appreciated. At the same time, progress in biotechnology led to linkage of the seed industry with the chemical industry, and to a series of mergers that have concentrated a major part of the world's seed production into the hands of a small number of transnational corporations. We have reviewed developments up to the end of the 1990s elsewhere (Brookfield 2001), and there is no space here to renew discussion of so large a topic. Nonetheless, something does need to be said concerning both the actual and potential effects on family-scale farming.

With all crops that are grown from seed, farmers have habitually saved seed from the preceding crop. Additionally it has long been the practice to obtain a proportion of seeds from elsewhere. This is a means by which genotype/environment inter-action can be tested and tolerance broadened, and yield improved. It is also curiosity; anyone who has travelled with farmers will have witnessed their eagerness to acquire new germplasm of any kind to try at home. Farmers have constantly modified natural processes of seed development by selection and cultivation. Their informal systems for acquiring, selecting and propagating seed became a topic of major interest, with an important literature (e.g. Zimmerer 1996; Brush 1999; Almekinders and Louwaars 1999; Brush *et al.* 2003). The development of modern crop breeding in the nineteenth century and its great flourishing in the twentieth century have created a new 'formal' seed system in which the suppliers have been either government agencies or private entrepreneurs. Although the formal system now provides most of the germplasm used in developed countries it does not supply all; in Europe and North America a significant proportion is still saved from previous crops (Almekinders and Louwaars 1999).

Hybrid seed cannot successfully be used in this way. Hybrids lose their charac-teristics in succeeding generations, and must be purchased new from the breeders every season. This first became of major commercial importance with maize in the USA and, with the import of hybrid maize to many developing countries, the annual purchase of new seed is added to farmers' costs. Unfortunately, the seed-chemical industry has become very keen for farmers to purchase all their seed, especially so because the cost of breeding new varieties, with or without transgenic modification, has grown higher and higher. Patents are now taken out in all possible countries to protect new seeds (and some seeds that are scarcely new at all, but have been captured by the industry). There is also a network of international agreements to protect these patents, as with proprietary medicines for human disease.

In developed countries the companies can achieve protection of their patents. A system not unlike policing is used to determine whether, for example, varieties genetically engineered to resist proprietary herbicides have been 'illegally' planted. Many farmers have been taken to court. 'Policing' is costly for the sellers of seed as well as the buyers, so that the industry grasped with enthusiasm at what were called 'genetic use restriction technologies' when they were first patented in 1998. Principal among these is breeding genetic traits into the seed that will not allow second generation seed to germinate without application of a proprietary chemical. This 'terminator' gene has attracted fierce opposition from farmer and civil society organizations, and for this reason it still remains only a threat, though a real one. It is obviously in the interest of the seed-chemical companies that farmers buy all their seed from them, as well as the chemicals needed for them to survive and grow. The debate drags on and many governments have been convinced by the company argument that only in this way can the high yielding seed needed to meet world food requirements be economically produced. As with livestock, each new development turns the screw on the farmer by an additional, and costly, twist.

Notes

1 There are striking historical examples. In France, in the century after the Black Death of 1348 it became increasingly common for large extended families, multi-generational or unions of brothers, to form under very tight conditions of mutual obligation to cultivate the larger areas made available by depopulation. The only alternative, other than leaving land unused, was to employ labour in a market where the labourer could command a high wage. These arrangements faded away once population growth resumed in the sixteenth century (Ladurie 1976).

2 For teenage children there are often rates of pay and other conditions specified by legislation in the developed countries, but these are usually applied only where teenagers work for farmers other than their own families, specifically as casual labour or labour recruited by gangmasters.

3 Although the term 'farm labour contractor' is quite widespread, in different languages, and is gender-neutral, there is scope for confusion with other types of contractors. For this reason, and because the British term 'gangmaster' best conveys the nature of the social relationship, we use 'gangmaster' for preference in this book.

4 These outbreaks included not only foot-and-mouth disease, infectious to other livestock, but also the more dangerous bovine spongiform encephalopathy (BSE) and, thus far without effect on trade, highly pathogenic avian influenza.

5 From the farm to the consumer

Disposing of farm produce is as important as producing it. In this chapter we move away from the farm into the realms of trade, marketing and taxation, areas in which the farmer has to deal with forces outside agriculture. We begin with the off-farm disposal of produce in twentieth-century societies without money, then go on to discuss rural/urban market systems as they have evolved through more than a thousand years. This takes us then to some of the major changes of modern times, in which innovations in the commerce of food and fibre have succeeded the agrotechnical revolution of the past two centuries as a principal force having far-reaching consequences for the family-scale farm.

The social context of rural trade

The 'normal surplus' and its disposal

There are still many farms in the developing countries that provide most of their own needs in food and fuel, and some provide it all for sustained periods of time. There are fewer that make and repair their own tools and kitchen hardware and although some developing country farmhouses are devoid of even rudimentary furniture, most have some. Virtually none are without tools. Few farm families now make all or any of their own clothing and almost none can avoid paying some tax, rent or both. Many have to pay for labour, or at least feed the reciprocal labour of neighbours. They have to meet the costs of weddings, family celebrations and funerals, and many have to make socially obligated payments related to religion and public festivities. Farms have been elements in wider society and economy since prehistoric times, and a part of farm produce has always been used to facilitate social goals and, by so doing, enhance security in times of need.

To ensure their livelihood, and obtain the commodities and services that they need, subsistence-oriented farmers strive, for reasons of security, to produce what Allan (1965: 38) called a 'normal surplus', commonly between 10 and 25 per cent above estimated basic needs. This 'normal surplus' of subsistence-oriented farmers was first the product of risk management, but production of a surplus for disposal or sale arises from it. Viewed in the context of social history, the normal surplus has been the foundation of all trade, however institutionalized.

Social production: 'trade' through formal prestations

Individuals use their surplus for a range of purposes, but they are rarely allowed anything like total freedom of decision. If the surplus pays tax or a tribute, this fulfils an obligation. If part of the surplus is simply presented to affines or other relatives, or unrelated exchange partners, this either repays an old obligation or creates a new one. Such prestations then lead to other transactions, in kind or in services. For an organized community, the surplus produced by its members constituted a resource which could be mobilized for community purposes, or for the purposes of the leader or ruler. The size of this 'social production' can be substantial enough to raise total production far above the levels required just to feed the producing population, yet unless it was formally traded it was little regarded before the 1960s (Brookfield 1969, 1972).

Reciprocal exchanges between partners and whole groups, organized by chiefs or 'big men', remained important in many developing country societies until very recently, for example in the western Pacific region where they were classically described by many anthropologists, and some government observers, in the first 60 years of the twentieth century. The most famous discussion concerns 'Kula Ring' prestation/trading among the islands off the eastern tip of New Guinea (Malinowski 1922). In this case, as in some others, the giving and receiving of valuables was the formal part of the exchanges, with a host of other transactions added.

In the populous Chimbu region of central New Guinea, an economy of competitive ritual exchanges first expanded in scale during the short colonial period (1935–75) then swiftly fell away (Brown and Brookfield 2005). Chimbu had a very old system of inter-group exchanges of nuts, fruit and other vegetables, excluding only the staple food, sweet potato. A great heap of such produce was assembled by individual contributions, and set up close to the boundary of group territory. Boastful speeches by the leader proclaiming group generosity were then followed by individual distributions to relatives and other exchange-partners in the receiving group. At some time during the past three hundred years these exchanges were supplemented by the ceremonial killing, butchering and cooking of pigs, as many as a thousand in a day, in the central plazas of temporary settlements that were built to accommodate the guests. The building of pig herds for these large-scale prestations occupied several years and preparation required large-scale and sustained organization and planning on the part of the donors on each occasion. The receivers would later become donors in their turn, after several years.

Both forms of prestation remained active in 1960, when Brookfield and Brown (1963) observed them. What we saw was organized reciprocal exchange, as defined by Polanyi *et al.* (1957). When money began to enter circulation, which in Chimbu happened in the 1940s, it quickly began to replace the old forms of valuation of goods and services. Until a generation before 1960, manufactured goods traded in these exchanges would have included salt and stone axe blades, as well as the durable large ornamental shells traded in from the distant coast. Hughes (1973) argued persuasively that valuables, and salt also, were not only used as a medium of exchange, but that a constant demand for them stimulated regular production of the consumables that were also transacted in these formal exchanges. In some parts of the world, including East Africa, salt was the earliest commodity traded, being produced in only a few places and widely demanded for livestock as well as people. In western Uganda, where there were no marketplaces until the colonial period, salt trading from the lakes in the dry region of the western rift valley was already important several hundred years earlier and in the nineteenth century it came to be associated with trade in other commodities (Good 1970).

Periodic markets

We were fortunate to work in Chimbu before the earlier transaction form began to die away, which it did rather quickly between 1965 and 1985. In most parts of the world such transitions took place centuries earlier. In Java, a system of market exchange was already in place before the ninth century AD. As reconstructed from contemporary inscriptions by Christie (2004), there were intersecting rings of periodic markets held in large villages on a five-day cycle. Reciprocal prestations still took place on religious and ceremonial occasions, but the sale of most crops and livestock took place in the markets, and transaction was in cash.

The Javanese market system was not greatly changed in the mid-twentieth century (Dewey 1962). Each market had a manager, who was responsible for collection of fees or tax. In addition to the numerous small traders mostly selling their own produce, were wholesale dealers (*bakul*) who bulked produce for onward transport to other markets and even other islands. The market circuit was followed by artisans and vendors of services, who were able to build up adequate business by meeting different customers each day. In Christian and Muslim countries markets were usually held every seventh day, but in other regions and in countries that had markets before either religion held sway, the periodicity of markets varied between four and eight or even sixteen days. The organizing principle of these market circuits was classically analysed in China by Skinner (1964). They spread marketing opportunity throughout a large rural area, and made possible the distribution of services over the same area. Similar market rings have been described in many countries, and they were until modern times widely present in Europe. Often the market rings developed independently of the administrative network.

Rural periodic markets were not necessary for commercial trade in rural areas. They have not flourished in any post-sixteenth century European settlement, although attempts to set them up have been made from time to time. They are only a colonial-era innovation in a large part of sub-Saharan Africa, except West Africa. Even organized urban retail markets have not developed everywhere or at all times, although wholesale markets have been almost universal. In lands of European settlement the fixed-point trade store was common, where the buyer of principal crops and the supplier of the farmers' regular needs was the same person. In western New York State in the early nineteenth century:

> A farmer in our period would drive up to the store with sixty bushels of wheat, for instance, in his wagon. After a dicker with the storekeeper, he would sell the grain and make purchases from the variegated stock of the general store. . . . No cash had changed hands; the storekeeper had simply set down the grain in his ledger on the credit side of the farmer's account and the variegated articles taken away on the debit side, which was generally the heavier of the two.
>
> (Albion 1939: 76)

The supply of wholesale markets is a specialized business in most countries. Produce is collected through stages, whether through periodic markets, by fixed-point traders or mobile collecting agents. Commonly on the retail side in developing countries, hawkers, carrying or wheeling their goods along streets and to houses, remain an accompaniment to marketing. Shops appeared early in many regions, combining trade in local produce with sale of manufactured and imported wares. In Europe, they began to displace market traders and producer sellers in the seventeenth and eighteenth centuries and massively so in the nineteenth century. Retailing of foodstuffs stayed in the markets for longer than retailing of clothing, artisan goods and manufactured produce, but by the early twentieth century the periodic produce markets in European countries were only peripheral to a retail economy that was overwhelmingly carried out in shops. Producer-sellers almost vanished; it is only with the fringe 'farmers' markets' of the most modern period that they have returned to the urban scene in Europe and North America.

Rural periodic markets in the developing countries have a large modern literature, usually discussing their character, distribution and modus operandi at a point in time. Most studies are based on surveys undertaken on from one to a few dozen market days. Rural markets are dominated by part-time producer-sellers, but almost all have a proportion of professional full-time sellers who deal in larger quantities and often also retail manufactured goods (e.g. Dewey 1962; Geertz 1965; Brookfield 1969; Good 1970; Alexander 1987). The part-time sellers are mainly farmers or their spouses. The goods and, sometimes, livestock that they sell come from their own farms, or are manufactured in the farmhouse. The part-time sellers are often also buyers, for varied produce is on sale, including cooked foods and farm tools of both local and foreign manufacture. Small markets are places where friends and acquaintances meet and where business of all kinds is transacted. Professional traders and providers of services such as barbers, dentists, letter writers, bicycle repairers, tailors and many others visit different small markets in succession.

Within a regional economy there is often a hierarchy of markets, with goods sold in one market then being resold in another where demand is greater. Complex personal links secure supply and provide credit, both within the marketplace and around it into the villages, and outward toward the major regional and urban markets. Markets are places in which peaceful conditions must be ensured to facilitate transaction, and the great majority is therefore licensed by government, which collects dues from the sellers and often provides buildings and other infrastructure. Many were founded by rulers, to institutionalize supply and trade in territories that they controlled.

Early in the literature on markets is the account of marketing in central Nigeria by Nadel (1942), based on his observations in the 1930s of a market in a village of 3,000 people, probably set up in the 1840s. In the emirate of Nupe, discussed in Chapter 2, every village had its periodic market. The larger ones had permanent buildings, booths and stalls. Kutigi market operated in a small way every day, but principally operated on the designated weekly market day when it attracted from 400 to 500 traders.

> Every kind of trade or craft is represented even in the smaller village markets. In the roofed booths which every market possesses the more valuable goods are displayed – products of native industries as well as European goods. Tailors and leather workers work in open huts. In the open square framed by these booths and huts agricultural produce, food-stuffs of every kind, live stock, pots and tools are offered for sale. In one corner of the market barbers put up their open-air shop. In another corner butchers sell their goods, a drummer endlessly repeating the same rhythmical motif – the signal that the butchers have slaughtered that day – to attract the attention of the crowd.
>
> (Nadel 1942: 322–3)

In a region where there was widespread understanding of variable supply and of expected prices, there was little or no haggling except for articles of high value. Buyers, and sellers too, came for the sake of company and entertainment as well as trade. Entertainers were often present. Local attendance at village markets came from within a radius of 3 to 6 km, the area covered by local knowledge and interest in each others' affairs, and within which most intermarriage between villagers took place.

Supplying the cities

Markets have always been dynamic, growing, increasing in number and frequency, or falling away into disuse with time. Periodic markets can become daily markets, and then can fade away altogether as trade is captured by other markets or other agencies. Amanor (1994) described how the competition between marketplaces in southeastern Ghana for the trade in food for the capital city led to some rapid changes. Close to growing urban areas, rural markets tend to be swallowed by the city markets, in part as dealers intercept supplies or livestock before they can reach the market and sell them themselves in the city. This is a very old practice; it was complained of around London in the 1790s (Middleton 1798). It was a major feature in Bihar, India, in the 1970s, where small markets near the industrial city of Jamshedpur were stifled in this way (Wanmali 1981).

Among the many modern accounts, one of the more informative is of Java at a late stage in the rapid modern commercialization of the rural economy (Alexander 1987; Alexander and Booth 1992). Basic collection of produce in small lots, mainly by women, supplied not only market stallholders but also wholesalers, the *bakul*. The *bakul* additionally employed agents to collect produce in larger quantities, often using small trucks and supplying the drivers with funds for purchase in the villages. These wholesalers were, as centuries earlier, the pivot of the system, now both bulking rural produce for onward transmission to the city markets, and supplying factory produce to market stallholders and rural shops for local sale. But by the 1980s depot operators and government-supported cooperatives, using the improved infrastructure and transport system, were able to do their own bulking, thus eliminating the smaller intermediaries. At the same time, urban manufacturers were becoming better attuned to the marketing needs of the lower-income rural customers, producing small-pack goods in volume to be sold at affordable prices, often bypassing the market altogether and selling through the growing number of fixed shops. In Indonesia as a whole there were about five million unincorporated trading enterprises in 1986, employing almost nine million people in the countryside and the cities. But the small market traders and urban hawkers were seen by Alexander and Booth (1992) to face gradual extinction. If the model of the manner in which shops and later supermarkets had largely replaced urban retail markets in western Europe between 1780 and 1980 is applicable, this was a reasonable prediction.

Madagascar

The central highlands of Madagascar are a land of markets to this day. Every small town has its market, and informal roadside trading is a common sight. Regulation began in the 1780s when the victor in a series of wars among the local groups into which the Merina people were divided, created the Imerina state and organized it as the base for conquest of the whole island during the next 50 years. His decrees involved reorganization of land tenure, improvement of irrigated rice farming, management of both free and slave labour and the creation of an army. Because his innovations entailed a greatly enhanced division of labour, a thorough restructuring of the already existing system of markets was necessary, and was imposed.

By the 1960s, Imerina's hilly capital, Antananarivo, had become a city of over 300,000 people. A large part of its trade was organized through a single central market, by that time concentrated in Analakely, a lowland area of the city close to the railway station on what had been rice fields until the late nineteenth century. The Analakely market is best known as the Zoma, a term derived from the word for Friday when the main market was held with more than five times the business done on other days. But most of the market was in daily operation. Gérald Donque (1965–66) detailed both the operation of the market and its supply with nationally produced and imported produce.[1] The city authorities had decided in the early 1960s to break up the overcrowded and insanitary central market into a number of local markets, and into shops. In the 1970s a major part of food marketing in Madagascar was taken over by parastatal bodies, but private marketing survived the socialist period. Now that Antananarivo has reached a population of over one million, dispersal of the central market has taken place, although its role in providing the city with fresh fruit and vegetables remains.

Donque (1966) provided intimate detail on the manner in which market supply and trade operated, and particularly on the role of intermediaries, professional traders who were successful in inserting themselves between the producers of most commodities and their ultimate retail customers. Independently, Bonnemaison (1967) provided valuable complementary detail about what happened to sellers in the supply area in the same years. Their material underlines the fact that once a farm becomes involved in trading its produce other than by direct sale to a consumer, it becomes involved with traders whose primary interest is neither in improving returns to the producer nor in reducing costs for the consumer; the traders' interest is in the maximization of their own profits. This may happen without the traders being, or becoming, capitalists. Even small-scale traders 'appropriate the surplus' from farmers and can do so on a substantial scale.

With rice, one of the largest single commodities sold in the market, separation of supplier from consumer was almost complete despite the fact that Antananarivo lies in the middle of a long-established rice growing region. Only limited quantities, being the normal surplus, were offered for sale by the farmers. Almost all marketed rice is milled, but the millers did not buy the rice directly. The country storekeepers were the main intermediaries, and they employed collectors who advanced money to the farmers at high interest rates; the debt was recorded as unmilled rice to be delivered at harvest. The collectors were indebted to the storekeeper who was in turn indebted to one of the numerous rice mills. After milling, the rice that reached Antananarivo went to 20 or so urban wholesalers, who in turn resold it to retailers with permanent stalls in the market. Other rice was stored and went back into the rural marketing system during the pre-harvest season when shortages were widespread (Donque 1966).

Other dry foods, maize and potatoes in particular, were handled in a similar way. In a region about 100 km south of Antananarivo, potatoes were the principal crop sold. Early potatoes grown on natural levees above the rice fields sold well (Bonnemaison 1967). Most potatoes, however, were grown in higher areas around 2,000 m. They were brought down to the regional town, Ambohibary, by ox-cart in a six-hour journey, and were then offered to one of a dozen intermediaries who forwarded them to the wholesale merchants at Isotry in Antananarivo. The intermediaries drove hard bargains:

> The merchant states his price, and there is no arguing with him. . . . He then does a triage on the potatoes, and [often] a quarter or a third of the load is declared unsaleable. The crop bought is then weighed and put in 25 kg bags, the buyer taking a fee of 2 kg on each bag for his own expenses. . . . He then proposes to the peasant that he buy at half-price that part of the offering that he has rejected. [Rather than carry the potatoes back home] the peasant accepts,

but [it can be supposed that] most of the 'unsaleable' potatoes are mixed with the others for dispatch to Isotry.

(Bonnemaison 1967: 217; translation by Brookfield)

By contrast, the three rice merchants at Ambohibary themselves visited the fields around the town to arrange deliveries from the farmers. They developed close working relations, lending money to the farmers and additionally selling them milled rice in the lean season before the next harvest (Bonnemaison 1967).

Fruit and green vegetables occupied a major area in the Analakely market. Most were produced within 120 km of Antananarivo, and a much larger share of production was grown specifically for market sale than with rice. Most producers grew vegetables in the rice fields during the dry winter and on colluvial slopes above the rice fields in summer. The growers travelled on foot, by country minibus or by train to deliver to Antananarivo, but few of them got to sell their goods in the market. Most were intercepted on their way into the city, or at the bus and railway stations or within the market itself, by intermediaries who put heavy pressure on them to sell their produce on the spot. The big selling day was Friday, and most of the supplies came in on Thursday. The intermediaries would either be retailers themselves, or would sell to retailers, often spending the night under blankets in the market area while one of their group kept watch against thieves. At the end of the 1950s the authorities built a shelter where bona-fide producer-sellers could spend the night, but in the early 1960s few were using this as they needed to return home to work on the farm. It was even to their advantage to do so. Donque provided an example:

Take this farm lady from near Antsirabe who wished to sell three baskets of peas and one of onions. She spent 400 francs to reach Antananarivo by train, and another 200 to return home. She spent 400 on a meal in town. Selling her 20 kg of onions and 45 kg of peas to an intermediary gained her 3250 francs, a net gain of 2250 francs for a journey of almost 300 km [*at that time the Malagasy franc was worth 0.02 French francs, so she gained about 45 French francs or about US\$7.50, as total return for production and sale expenses*]. This was little enough, she agreed, but added that had she taken her produce to the local market and sold it herself, it would have taken her three times as long, and yielded a lower price for the same merchandise.

(Donque 1966: 170; translation by Brookfield)

The common pattern with the majority of commodities was that wholesale trading in the Analakely market, or of larger quantities at the premises of wholesalers, would continue right up to the time when a bell was rung to declare the market open. Then everything was between the retailers and the buyers. Asking prices would exceed those offered to the producers by from 1.6 to 2.2 times, but it was normal to bargain for a considerable time before an agreed price was reached, so that the final mark-up was commonly reduced by 20–30 per cent. The weekday market had about 1,200 stalls, of which 550 sold food. On Fridays the total increased to 3,300, of which 1,300 were selling food. Most stalls were staffed only by family members. None but a few who held permanent stands mainly selling cloth and clothing would employ assistants, or could command any credit with banks or merchants to expand their businesses. Most stallholders operated on rolling debt to their wholesaler and intermediary suppliers and other trading partners, commonly small-scale Chinese and Indian merchants. Almost all stalls were held by people who had no other profession but trade. While the small intermediaries might exploit, threaten or even swindle the producers, the system almost totally constrained their freedom of action, and trading brought them only a small income.

If the cascade of intermediaries was massive in the supply of grains, fruit and vegetables, it became far more so in regard to meat, for most of Antananarivo's meat supply came not from the immediate vicinity but from the western and southern parts of the large island. Livestock, principally cattle, were first sold in local markets at prices that varied greatly with the season and the state of the natural pasture. They were then driven over a period of weeks, and resold at market after market and sometimes (illegally) on the cattle track. In the final stages, the last 100 km or so into Antananarivo, some more promising animals might be bought for fattening, usually on hay that was also traded in the market, for there was nothing in the way of improved pasture. These might then go to private butchers in the city. The majority of animals would eventually reach the municipal abattoir situated by a river west of the city and be slaughtered. The meat still had to pass through intermediaries on its way to the meat sellers, who were mostly permanent stall-holders. On a per kg basis, the final retail price of meat was commonly over three times higher in the urban market than in the rural areas.

Analakely market was bordered by densely built slopes on which numerous services to the market were located, including wholesalers, depots for the storage of unsold goods, suppliers of cheap stand-up meals and restaurants of somewhat higher quality, cheap hotels providing overnight lodging under crowded conditions, and some shops. With an almost total lack of toilet facilities other than open trenches, the whole market area was not only crowded but very insanitary, and in the wet season great numbers of large flies infested the fresh produce sold in the market. Thieves and pickpockets were a serious menace. Already in the 1960s the main road running past the market was becoming the site of large new shops and offices, including one supermarket. By the 1980s, this had become part of the modern commercial heart of the city. The market had become a nuisance to the entre-preneurs, and a dangerous place for the unwary. Other market areas were created in different parts of the city, and supermarkets became numerous, obtaining their supplies from the wholesalers and probably by direct supply chains contracting rural wholesalers. Nonetheless, two areas of the Analakely market survived still in 2005, carrying on the same range of business. In the smaller towns of Madagascar, shops remain, as ever, peripheral to the marketplaces.

West Africa

Marketplaces in West Africa are even more deeply embedded in society and history than they are in Madagascar. Mabogunje (1968) traced their origin from long distance trade between the Atlantic coast and North Africa, extending back almost a thousand years. We have already mentioned the active functioning of a Nigerian rural market in the 1930s. In Ghana, the cities of Accra and Kumasi derived supply from a wide area, and growers in the north supplied yams and cattle to intermediary buyers who travelled the roads. Vegetables, manioc (cassava) and other produce were the main commercial speciality of villagers closer to the two cities. In Nigeria, every small pre-colonial state had its royal residence and market (Mabogunje 1967). The urban markets were supplied by rural periodic markets in the surrounding countryside, as well as by trading chains carrying produce over long distances, such as cattle from the north to the cities in the south, and kola nuts, valued as stimulants in the alcohol-free Muslim regions of the north (Hodder 1967).

The fullest discussion available is on the markets of Ibadan, then capital of the Western Province of Nigeria (Hodder 1967; Mabogunje 1968; Guyer 1997). Ibadan was established only during the nineteenth century, after a series of wars which destroyed most other towns in the Yoruba region. Until the 1960s, Ibadan was the largest city in inter-tropical Africa, and its food supply was a principal activity for farmers of a large region, in addition to cocoa and palm oil that were produced for export. Its market is described in Box 5.1.

Supply to the large cities of southern Nigeria now draws not only on Nigerian farmers but also on farmers and pastoralists in Niger, to the north. Niger farmers began selling livestock, through two or three intermediaries, to forwarders who set up when the railway reached Kano, near the border, in 1911. Use of road transport more recently has enabled market traders to set up further north, beyond the limit of agriculture, and the pastoralists have made full use of these opportunities (Kerven 1992). Nigeria's success stands in the face of so many discussions of the so-called African agrarian crisis, in which the indigenous market system is often described as inefficient, fragmented, imperfect and inadequate! Yet farmers and pastoralists have shown themselves very responsive to market opportunities, and traders have been equally ready to take advantage of new openings. Ibadan, and now much larger Lagos to the southwest, plus Kano in the north, continue to be adequately supplied through an innovative production and trading expansion that has characterized most of West Africa.

Credit in the marketing system

Most retail fresh produce trade takes place in cash, but individual retail traders, as distinct from the supermarkets, frequently offer credit to customers. In Yoruba, migrant labour was mainly paid only at the end of the season, and credit channels were extensively employed with not much support from either banks or larger traders. Supply on credit of produce and of services has characterized many rural market systems, which operate without large cash reserves. The cash flow itself fuels the working of the system, and this therefore demands widespread trust, and ready information concerning the rare defaulters. Most market traders have limited capital, and lend on the basis that they know that the debtor will repay, at least in

---

BOX 5.1 THE MARKET AT IBADAN, NIGERIA, AND ITS SUPPLY

In the 1960s Ibadan had a very crowded central market, containing both wholesalers and retailers, held in the old streets of the central city. There was a more modern market near the railway and bus stations, and the livestock market had been moved to the northern outskirts, intercepting the main inward flow of cattle from the north. Morning and day markets mainly dealt in fresh produce, much of which was bought by trading women who then prepared cooked food for sale in the several night markets in which most Yoruba people in the city obtained their evening meals (Hodder 1967). Although trade through shops was increasing, up till then it had not reduced market business. Ibadan markets also served as the source of wholesale supply to many lesser urban and rural markets in Yorubaland and beyond. Subsequently the city has grown from under one million to more than three million people. Its markets have not died, and even some of the supermarket opposition has had to close down.

The market area has responded to the rising city demand, not so much by expanding as by internal reorganization of production and specialization. Jane Guyer (1997) analysed the changes in a western Yoruba community within Ibarapa district, about 100 km from the city. Until the 1960s its main contribution was the supply of dried melon seeds, commanding a relatively high price with low bulk. Buying markets then developed in the rural parts of Ibarapa, leading to trade in cassava flour which became the main cash product of small farmers and especially of the women farmers who emerged alongside the male farmers in the 1970s. Trucks reached these markets to carry produce to Ibadan. Farm enterprise depended heavily on migrant labour. During the Nigerian oil boom there was major new development in the use of tractors, while Japanese-made multi-purpose vehicles extended the reach of travelling buyers. Latterly these have gone directly to the supplying villages, bypassing the local buying markets.

By the late 1980s, the boom was over; vehicles were rarely replaced and were kept in service with increasing difficulty. Nonetheless, the differentiation of producers and traders continued, and production continued to expand and diversify. The structural change was not capitalism, and some large-scale private farms did poorly, but consisted of a multiplication of the organized production and service niches which had always characterized the Yoruba economic system.

---

part, or ultimately. Money wealth is not measured so much by cash in hand as by the amount of credit loaned out in many parts of Africa.

As Ward (1960) effectively showed, this combination of small capital and small loans to a necessarily limited number of people helps explain the multiplicity of creditors in many systems, not only in market trade but among the fixed-point buyer-seller shopkeepers. In the absence of a system of banking that can operate at the scale of most rural trade, there are practical limits to the number of clients that any one creditor can serve. It is the minority of large and powerful creditors who operate specifically as moneylenders at high rates of interest who are the ones that have earned such obloquy over centuries. The mass of small creditors provides an essential service necessary to keep the system flowing, and they make only limited profits.

Wholesaling always involves credit, especially where one wholesaler supplies another or forwards goods to distant places. Managing such credit demands personal relationships of an unusual order or an institutionalized system for the certification of good clients and the penalization of defaulters. Platteau (1994) was mainly concerned with long distance trade on a large scale when arguing for the importance of considering 'real-society' institutions that must necessarily underlie any 'free' market. Nonetheless, his arguments have force even at the small scale of rural trading. Good pricing and trust based on familiarity and reputation remain central to the effective operation of fragmented markets. But they do make them work.

Supply chains and modern innovations

The provisioning of Antananarivo and Ibadan with meat involved long supply chains; both markets were stages in chains which supplied other markets – on the east coast of Madagascar in the Antananarivo case, and both in coastal and northern Nigeria in the case of Ibadan. The *bakul* of small town Java were links in the chains supplying the major cities of Jakarta and Surabaya. Supply chains are ancient, but have changed greatly in nature in modern times. Wholesaling intermediaries were

---

BOX 5.2 A SMALL INTER-ISLAND SUPPLY CHAIN IN THE WEST INDIES

The large urban market of Port-of-Spain in Trinidad has long depended on the smaller islands to the north for food supplies above what could be produced on the sugar-specializing island. Working on St Vincent in the early 1970s, Brookfield and a Vincentian student visited farmers who supplied the Trinidad market. One had his own transport to take deliveries down to the port and, in addition to sending his own produce, he bought from a number of neighbouring farmers, or forwarded their produce on commission. His wife worked as a wholesaler in Trinidad, and one evening while we were talking to him in his well-furnished farmhouse she telephoned to tell him the state of the Trinidad market, and what should be sent on the next ship to make a good sale. Although we could not hear her end of the conversation, it was obvious that the lady managed the business.

among the first to make effective use of telecommunications to improve the efficiency and competitiveness of their businesses (Box 5.2).

A wholesaler was in the strongest position to develop a small but enduring supply chain between scattered small producers and numerous small buyers. Even if producers and consumers form cooperatives to increase their bargaining strength, in any trading over distance there are always points at which the strands come together and can be controlled. Operation of a larger supply chain required capital, and, because of the dispersed nature of rural production, control was established mainly from near the retailing end of the chain, or from enterprises engaged in food processing and packaging, or in forwarding in bulk to other markets. Security of supply demanded contract arrangements (unwritten or written), often reinforced or created by indebtedness. Real control of large and complex supply chains arose among companies which processed, packed or ultimately sold farm produce rather than among the competing wholesalers. Brewing, using barley, was an early and continuing case in point. Taking account also of the tax which government could most readily levy on the beer, farmers received only 24 per cent of the final price of the product in the vicinity of London in the 1790s; London then supported 12 breweries (Middleton 1798). In southern Peru after 1950, a beer company based in Cuzco contracted barley production from a wide area, supervising its production through regional agents (Zimmerer 1996). At prices determined by the company, farmers received seed and fertilizer and sold all barley to the monopoly buyer. Many other commodities have been caught up in supply chains in modern times. Bottling, canning and sauce processing of fruit and vegetables became big business in the mid-nineteenth century, especially in the USA where the commanding role of food processors in supply chains has lasted. Contract farming in the USA was already well established in the 1960s, requiring contracting farmers to produce for the requirements of canneries, packers and food wholesalers. In Europe, and elsewhere, the controllers of modern supply chains have more often been the larger retail chains. Also important and still expanding are what are called 'foodservice' outlets, institutions, hotels and restaurants including the large and highly concentrated fast-food companies which have become important buyers with supply chains of their own.

To the processor, the large retailer and fast-food chain, the reasons for setting up a supply chain to bypass the rural and urban markets, and the wholesalers who operate within them, are clear. The old market system matches supply with demand at a moment in time; it cannot ensure continuous supply at a predetermined rate. Price fluctuation is endemic in the daily efforts to clear the market. Nor is quality control easy to achieve. First processors, then retailers, have sought to manage these problems by engaging suppliers under a very wide range of contract arrangements. To be sure, the chain to the processor is only a part of the whole chain from producer to consumer, but the locus of control is important. Whether chains are set up and dominated by freezers, canners, packers, manufacturers of soups and sauces, and other processors, or whether the chain is set up by final retailers, is of considerable significance from several points of view among which a spatial one is of relevance.

Mexico

Whereas major retailers are necessarily located centrally to the consuming population, food-processing firms have set up largely in relation to the location of their sources of supply. Thus processors sometimes collect produce from contracting farmers, or receive produce from farmers' own vehicles. An example of chains developed in this way is provided by the central Mexican state of Guanajuato, a principal supplier of frozen vegetables to the markets of the eastern USA (Echánove 2001; Echánove and Steffen 2003, 2005). Two canneries were set up in the early 1960s, and a vegetable freezing plant set up in 1967 has since been joined by nine others belonging to American and Mexican companies. All sell in bulk principally to the American market and contract their supplies from local farmers, mainly the large private farmers and only to a limited extent from the small farmers. The processing plants are close to the producing areas but are also strung along a belt close to the Pan-American highway which is the main route of dispatch. Companies select growers on the basis of ability to supply the required vegetables at a satisfactory level of quality. They must be financially solvent, have access to irrigation and machinery as well as transport in order to deliver their produce to the factory according to the agreed schedule. Companies provide seed and some principal agricultural inputs, together with the sort of technological assistance that was traditionally provided by agricultural extension services. Farmers gain a more secure market, and in this competitive situation some are able to reduce their risks by contracting with more than one factory and also growing produce for sale fresh in the open market. Onward supply to large-scale buyers in supermarkets and foodservice outlets in the USA is a matter of direct contract between the processor and the final bulk buyer. How the small farmers manage in this system is described in Box 5.3.

Dominance of the supermarkets

The driving forces of farm trade reorganization were initially food processing and then managed national and international distribution especially under the dislocating conditions of 1930–50. Since the 1960s the drivers have shifted to the development of fast-food chains and, most decisively, the growth of supermarkets. Governments have assisted, for example in France in the 1960s, but fast-food production and retailing represent an area that has received minimum government intervention everywhere.

The control of a retail-focused supply chain serving city markets is inevitably further removed from the farmer, and very commonly intermediaries find an important niche in the system. Given the variability and short life of most fresh produce, quality control is centrally important, and requires strict systems of grading and inspection. It also requires ways to ensure that supplies reach the retailers in a timely and smooth pattern, minimally affected by gluts and shortages. Even at a considerable distance from the final bulk buyer, just-in-time deliveries to the contracted intermediary are necessary in order that the latter can provide the service

BOX 5.3 SMALL FARMERS CONTRACTING WITH A
SUPPLY CHAIN IN MEXICO

Few of the contracted farmers in the Mexican state of Guanajuato are small farmers, the land reform *ejidatarios* described in Chapter 11. Echánove (2001) provides interesting information on what happens within a community in which there are some contracted *ejidatarios* supplying vegetables to the freezing and canning companies. Among 32 *ejidatarios*, six, plus the legally landless son of another, have for some years grown broccoli and cauliflower under contract for the freezing companies, as well as other vegetables and maize. As financed contractors, they receive seedlings, fertilizer, pesticides and technical advice – plus supervision – from the company's extension workers. The full costs are deducted from their regular payments. They have to collect their inputs from and deliver their crop to the factory. To remain as contractors, they have had to grow larger in order to remain on the books and compete for additional contracts. The companies no longer need to seek their business, preferring large farms so as to reduce their own transaction costs. Reduction in the number of contractors and enlargement of the average size of quota fields are common features of 'mature' contracting schemes, which in the initial stages had to cast their nets more widely.

The farmers have expanded by renting land from other *ejidatarios* in their own and adjacent villages, and entering into sharecropping arrangements to obtain larger areas which they can offer for contract, and ensure well-water for irrigation. While village plots average only 5 ha, some of the contractors cultivate 25 ha or more and employ labour on a substantial scale because broccoli requires 80–100 working days per hectare in each growing cycle of three months. Successful growers have been able to acquire vehicles and machinery, expand their production through the year and improve their houses. Many were still contracted to the freezer companies six years later (Flavia Echánove, pers. comm. 2006).

This expansion entails the 'reverse tenancy' introduced in Chapter 3, in which larger farmers rent from the small. The latter get rent from their small plots, and also income as workers on them. The same has happened on a larger scale in the Indian Punjab, where the number of true smallholders has been significantly reduced in consequence (Singh 2002). Contracting does not turn the successful contractor into a semi-proletarian, but it does so to those onto whose land he is able to expand. Yet in the Mexican case no one seemed to mind as there was more money and work in the local economy. The main hope is that the good times will continue, despite increasing competition at all levels in a tightening global market.

the retailer requires. Since the advent of air freight on a large scale, chains have become fully international, so that year-round supplies can be ensured even across the hemispheres. Fresh fruit have been sourced in Chile and South Africa for the winter western European market, to some degree supplanting sources in North Africa since the 1980s. In the same period, the supply zone of fresh vegetables for the North American market has been expanded from Florida, California and Mexico to the countries of central America, where these 'non-traditional agricultural exports' have become important business for contracted small farmers. Most produce sourced in these ways reaches supermarkets without passing through the wholesale markets of the open system.

The supermarket had a nineteenth- and early twentieth-century ancestor in the form of department stores and bazaar-type stores retailing clothing and other dry goods. They moved into foodstuffs first selling dry foods, grains, roots, noodles and processed and packaged food. Fresh and perishable produce remained longest in the domain of the rural and urban open markets. Even in France, where supermarkets and hypermarkets have roughly 70 per cent of all food retailing, about 50 per cent of fresh fruit and vegetables was still handled outside the supermarket system at the beginning of the twenty-first century (Reardon *et al.* 2003). The same pattern emerges in the rapid growth of supermarket trading in developing countries. Whether the new pattern of retailing remains in the hands of local business or has been acquired by multinational companies, trade in fresh produce has resisted restructuring for longest.

Development of reliable supply chains is very demanding of management skills largely because of the volatility of both supply and demand. Imperfect information makes it very difficult to replace the old market system even in developed countries where information flow can best be managed by use of modern technology. Inventory control is not an available strategy with perishable goods and retailers must ensure regular supply. Suppliers need to clear their whole crop or livestock output, albeit at discounted prices, so that there remains a place for wholesale and retail markets even in the presence of the most efficiently managed supply chains (O'Keefe 2005).

In the new mix of developing country trading, the supermarkets in some countries already approach the 70–80 per cent share of total food retailing characteristic of North America and the countries of western Europe. They serve a large part of the population, poor as well as rich. The evidence of success for supermarket operation in parts of Latin America and Asia demonstrates fairly clearly that there are major economies to be made through efficient supply-chain management. Chains can quickly be developed with satisfaction to enough producers who accept the demanding contracts that are necessarily imposed. Convenience and speed of shopping are principal attractions to the customer. However, in some Latin American and Southeast Asian cities, observation suggests that people use supermarkets where access to fresh produce markets on a daily basis is sometimes difficult; those without refrigerators have to buy almost every day. The amount of fresh produce going through supermarkets is small compared with other goods.

Substantial financial resources are needed to develop and sustain supermarket systems and often these have been provided by the large multinational companies that have acquired a fast-growing share of the retail business in many developing countries. By doing so, the multinationals have been able to find new areas of operation which offer larger profit margins than are possible under the intense competition for an unexpanding food market in Europe, or a slowly growing one in North America. The French chain Carrefour, for example, has gained margins that are three times larger in its Argentine stores than in France (Reardon *et al.* 2003). It is envisaged that a very small number of food and general merchandise retailers might come to dominate business globally in a decade or so (Wrigley 2001). Many writers, from Goodman and Redclift (1981) onward, have found in agribusiness the modern means by which capitalism has 'appropriated the surplus' from the small farmers, replacing the landowning capitalist as the main force of control and accumulation. Writing in the early 1970s, W. Smith (1973) offered a more nuanced view, showing how an enterprising distillery in Quebec could provide a major support to farmers by contracting their grain. This dual aspect of contracting has continued as the system has grown and become more widespread.

Contract farming

Contract farming is essential to the development and management of supply chains. It has attracted a great deal of comment in a large literature. Contracting is not a new phenomenon in agricultural marketing, since it has been present in most share-cropping arrangements for hundreds of years. Asano-Tamanoi (1988) effectively demonstrated the continuities between the old and new arrangements in Spain and Japan. The contracting cooperatives that grew to dominate Danish dairy and pig production in the early twentieth century were the focal points of emerging European supply chains in these products. By the end of the twentieth century, half or more of all farm produce was being produced and sold on contract in both North America and Europe. Since the late 1980s there has been very rapid expansion of contracting arrangements along supply chains in Latin America and Southeast Asia, and most recently in Africa (Reardon *et al.* 2003). They have become integral also to commercial relations between developed and developing countries. Many of the companies now engaging in contract arrangements in the developing countries are either multinationals, or sell to the developed countries.

Management specialists and World Bank economists tend to regard contract marketing, and other forms of vertical integration, as a 'higher' form of market organization, and encourage its extension. The integration lays stress on efficiency, and offers more secure incomes to those farmers who can meet the contract requirements. Unless they have had bad experiences with rapacious companies, as has happened to some fruit farmers in Chile (Murray 2002), many farmers are happy to be contracted. Cook (2005) sees opportunities for Australian producers in contracting with supermarket buyers if the producers can link themselves successfully to, or combine in the creation of, an effective intermediary shipper. Farmers may not agree. They are forced into it as the supermarkets have been ruthless in increasing market share. Deregulation of the dairy industry has been good for those able to take advantage with contracts, but the margins are still outrageously low for the farmers, and supermarkets try to drive prices lower.

Many writers have seen great dangers for the independence of farmers. From the time of Vogeler (1981) and Goodman and Redclift (1981), contracting – even if beneficial in immediate money terms – has been seen as proletarianization of the once independent family farmer (Watts 1994). Lacking control over his or her own production decisions, the contracted farmer is seen to become a self-employed semi-proletarian, exploiting his and his family's labour in a Chayanovian manner to meet the demands of the contract (Watts 1994). In another global presentation, Glover and Kusterer (1990) do not go so far, but draw attention to a wide range of potential disadvantages for the contracted farmers.

The great variation in contract terms defies generalization, and there is also great variation in the way that contract projects are managed. FAO sponsored a handbook on their management, especially in developing countries (Eaton and Shepherd 2001), which showed many reasons for discontent and, sometimes, failure. They offered some excellent advice on achieving smooth management, some of which mirrors advice offered from a close comparison of schemes in Nigeria and South Africa by Porter and Phillips-Howard (1997). But Eaton and Shepherd, like others on the 'official' side, see contracting as a partnership between sponsor and grower, and do not place any stress on sponsors' preferential selection of mainly large farmers for receipt or renewal of contracts. Studies both in Mexico (Echánove 2001) and the Indian Punjab (Singh 2002) show this much more clearly.

Although there are schemes on which tens of thousands of farmers work for one sponsor under contract, many if not most of these arise from resettlement or land reform schemes, or involve former workers on estates that have either ceased to exist or survive only as 'core' estates for which the contractors are 'out-growers'.[2] Where arrangements are newly set up by agribusiness among individual farmers, the numbers contracted are often quite small, seldom more than a few hundred unless a parastatal organization works in cooperation with the private entrepreneurs. Even while maintaining diversified farms, contracted farmers in the Indian Punjab were able to put more land under crops than the average total holding of farmers in the region (Singh 2002). True smallholders only rarely participate, notwithstanding the literature in support of contracting as a means of development and poverty relief.

In conclusion, it is worthwhile to note that contracting is essentially a formalization and enlargement of the old market system, as described above for Madagascar. Contractors have to risk rejection of their produce on 'quality' grounds that increasingly arise from the capacity of the factories and supermarkets, rather than from consumer discrimination. When this happens, their position is no better than that of the potato producers at Ambohibary (Madagascar), perhaps worse because perishable produce is hard to sell elsewhere and often has to be discarded. The same or related problems arise throughout the contract farming business in all commodities and in all lands. Globalization can bring improved incomes and welfare, but it can also hurt for reasons that have little to do with the efficiency of the producing farmers themselves. They are among those who suffer from being 'on the low end of a tilted global agricultural trading system' (Jayne *et al.* 2005: 1). There are many in this position.

Notes

1 Donque's book-length study was published in a Madagascar periodical of very limited international circulation that ended in the 1980s. Hence it has not achieved the notice that it deserves for its content and analysis, as well as its lively and engaging style. Brookfield visited Madagascar in 1967 and he obtained all available copies of the journal; few international libraries have it.
2 A striking example is in Fiji where some 20,000 contractors supply sugar cane to the four mills of the parastatal Fiji Sugar Corporation. Most are descendants of the indentured workers imported from India by the multinational corporation which established the industry in the 1880s. This corporation withdrew after a report that recommended a significant increase in the share paid to growers was adopted by the government.

6   Farmers and the state

The leading role of the North Atlantic countries

We now come specifically to the role of the state, and in this chapter open discussion on the far-reaching developments that took place in North America and Europe in the twentieth century. From our point of view, this is in many ways a sad chapter to write. While policies in support of agriculture have all been justified by a rhetoric in which the importance of family farming has been emphasized, the policies themselves have strongly favoured large farmers, and small family farmers have received at best only minor benefit. The whole period since the 1930s has been one of major technical revolution in agriculture in all parts of Europe and North America, beginning earlier in some countries and regions, and later in others. It was taking place everywhere by the late 1940s. This is when the main yield increases were experienced in field crops and livestock products. Grigg (1984: 14) offers good reason to conclude that everything that happened in European agriculture before the 1930s was evolutionary, and that revolution only began after that decade. At the start of massive state intervention, however, agriculture in this region and also in most others was in the deepest plight it had experienced in modern times.

Background: depression, the New Deal, self-sufficiency and war

In the few years between 1931 and 1940, everything changed in the relationship between farmers and the state in North America and Europe. The collapse of world-wide economic confidence in late 1929, and its sequel of deep depression in 1930–33, had a particularly devastating effect on farm economies. With shrinkage in both domestic and international markets, many farmers were unable to service mortgages or even sustain purchases of the already essential off-farm inputs needed for all commercial and semi-commercial farm operations. The problem was seen as one of production greater than the market could absorb. After a short-lived American attempt to stabilize domestic prices in 1930, the grain price fell to the lowest level since 1896 (Hacker and Kendrick 1949). In the USA, farm incomes declined by 50 per cent between 1929 and 1932. The crisis year world-wide was 1931. Britain abandoned its long-standing free-trade policy and imposed a range of protective import tariffs together with restrictive quotas, going on to create marketing boards to stabilize returns from a group of major farm products. Almost all other countries raised existing tariff barriers so as to protect their own producers. In Europe, unlike North America, the problem of reducing production was replaced at different times during the 1930s with the urgency of achieving greater self-sufficiency of food supply, at first to replace imports and later in response to the growing threat of war. Responses varied greatly between countries. France shifted tack rather swiftly between tariffs, minimum price support, subsidized exports and production controls (with subsidies for not producing) (Lamartine-Yates 1940; Tarrant 1980).

Germany set up a 'national food administration' (*Reichsnährstand*) in 1933, establishing planning control at all levels from producer to retailer, based around the inheritable but indivisible family farm (*erbhof*) of 7–125 ha. These comprised about half of all German farms and many were in financial trouble (Lamartine-Yates 1940). Results up to 1938 remained modest, but the support and control measures were continued through the war, and assisted in ensuring that until the last months of World War II (1939–45) the health and nutrition of the German people remained above any previously attained level (Lamartine-Yates 1960). An even greater achievement was in Britain – it imported more than half its food before 1939 and was greatly threatened by the partial, but deadly, German submarine blockade. Under firm direction, and with substantial subsidy especially for farm mechanization, food output doubled between 1938 and 1944 (Lamartine-Yates 1960). The improvement in diet and health of the British people was sustained right through the war and beyond.

In USA, Roosevelt's Agricultural Adjustment Act of 1933, and especially the definitive act of 1938, were the first to devise administratively effective means of subsidizing farmers' income. These involved direct payments to cover the difference between 'target prices' and actual market prices (deficiency payments), and loans to support production advanced against growing crops. Benefit went mainly to the larger producers of maize, wheat, cotton and tobacco. Smaller producers got little and departures from the land continued, further spurred by the improvement in urban employment opportunities that took place after 1938 and during and after World War II. The state had assumed a proactive role in agriculture which was not to be lost again, and persists to the present day notwithstanding all the subsequent changes in national economic ideology.

Through national food sovereignty to a new international regime

The wartime experience had lasting consequences. European governments determined to progress toward national self-sufficiency, especially during the lean years after the war. Under new legislation in 1947, the UK continued the wartime modernization subsidies and committed itself to pay farmers guaranteed prices, then after 1953 deficiency payments that would meet gaps between targeted and received prices. These involved direct payments to farmers and were not intended to affect consumer prices. In the same period the USA set minimum prices for selected commodities through the loan advanced to farmers each season. Farmers were free to relinquish their crops to the Commodity Credit Corporation (CCC), and most did so as the generous prices were above those obtainable in the market. The size of the stocks soon became an embarrassment, notwithstanding a set of measures designed to avoid overproduction, particularly the creation of a conservation reserve (set-aside land). The efforts using pre-1940 devices such as paying farmers not to produce crops that were in excessive supply also failed. Subsidized internal distribution helped but external disposal quickly became necessary. An important outlet arose through the massive aid for European reconstruction provided by the American 'Marshall Plan', an important part of which came as food, feedstuffs and agrochemicals. This was phased out in the early 1950s but after 1954, under the famous Public Law 480, this aid began to be shifted to selected developing countries (Friedmann 1993). While all these measures brought help to farmers, they did not raise farm incomes above the rapidly rising urban levels, so that the pre-war exodus from the land continued.

Unlike what had happened in the 20 years after the end of World War I in 1918, the economies of the North Atlantic countries continued to flourish without serious interruption for almost 30 years until the 1970s. In agriculture, rapid mechanization, the use of chemical fertilizers and pesticides, improved germplasm of the major crops, new methods of livestock management, the development of specialized machinery and the enlargement of fields were all major elements. They led to higher yields and production. The elasticity of supply of farm produce is very responsive to price changes, whereas domestic demand is much less elastic. The manner in which American and European policy instruments acted to support producers and protect customers in this period is clearly set out with the aid of simple diagrams by Tarrant (1980: 84–99).

Agreements in the 1950s allowed European governments to protect their growing arable and dairy farming sectors, even against American exports, while opening markets to American maize and soya as feedstuffs for the expanding European livestock industries. This convenient arrangement was enshrined in the formation of the common agricultural policy (CAP) of what was to become the European Union (EU) between 1957 and 1962.[1] It allowed European production to expand rapidly, while integrating the feedstuff and livestock economies of the Atlantic countries for a period. American food crop exports were directed toward Japan and the developing countries and in the latter these subsidized foodstuffs quickly commanded a leading share in the national urban markets to the detriment of national agriculture. Food dependency was created in a range of countries which had been largely self-sufficient before World War II (Friedmann 1993). It has persisted in most of them until today.

Internal consequences of the technical and policy revolution

Small farms received least benefit from the higher farm incomes that became possible in the North Atlantic countries. In particular, they received much less in subsidies, however these were calculated. Large numbers of smaller-scale farmers and farm workers moved into higher paid employment elsewhere in the economy, although at different times in different countries and regions. In the USA, farm numbers declined by 1.6 million in the 1950s and by 960,000 in the 1960s, after which the decline eased until the early 1980s. In Britain the slow 1850–1939 decline in numbers simply resumed. In much of continental Europe there was quick resumption of an inter-war rural emigration that had already reduced populations in areas such as southern Italy and upland areas of France and Germany. Once the urban and industrial economies had been rebuilt, the decline became massive.

In the 15 countries of the pre-2004 European Union, over 20 million farmers and farm workers left the land between 1950 and 1990 (Bureau 2003). Except in ecologically and economically marginal regions, there was no great decline in the agricultural area, but farms grew larger through purchases and amalgamations. A major part of the decline and disappearance of small farms in the 1950s and 1960s took place without great distress. Only after the greater economic convulsions of the 1970s and 1980s did the old problem of bankruptcy due to inability to service debts again become a major cause of farm closure in some countries, including the USA. An American glimpse on less traumatic change through a large part of the century is provided in Box 6.1.

BOX 6.1 THE TWENTIETH CENTURY ON ILLINOIS FARMS

Anthropologist Jane Adams (1988) traced the history of family farms in a hilly county in southern Illinois through most of the twentieth century. This was a region of maize or vegetable production and mixed farming. Until the 1950s, it was also one in which family farms continued to produce for their own subsistence as well as for the market. Most farms kept small herds of dairy cows and flocks of poultry, and women were active in marketing the produce in the local region. After the 1940s mechanization, which had hitherto been limited in the Illinois county, increased with readily available credit and farming became more heavily commoditized. Only 35 per cent of county farms owned tractors in 1945, but by 1964 they were owned on almost 90 per cent of farms. This, and other innovations, made it possible to dispense with most of the hired labour. On the basis of inadequate census data, the number of paid farm labourers in the county declined from 428 in 1940 to 70 in 1970. At the same time, competition from more fully capitalized farms in California (using Mexican wage labour) reduced the profitability of vegetable production. In addition, both dairying and chicken and egg production concentrated in specialized enterprises, both locally and beyond. Adams, whose study focuses on the decoupling of the household from the farm enterprise, notes that while some farm wives participated more fully in the work of the farm and its management, an increasing number found off-farm work. Farms declined in number, but up to the 1970s the now fully commercial family farm enterprise remained intact.

Innovations of other kinds

While a new global food system was being crafted internationally, a number of other changes were in progress. Some were technical, especially those leading to the Green Revolution in wheat and rice cultivation. Of major economic importance in the growing market for vegetable oils was the improvement through plant breeding of soya, and oil-seed rape as an alternative provider of vegetable oil in cooler climates, coupled with the increasing substitutability of vegetable oils by industrial research.[2] High fructose corn syrup from maize became a substitute for sugar in processed foods.

Significant changes were emerging in the organization of business around farming. Agribusiness, discussed in Chapter 5, grew first and most rapidly in the USA (Davis and Goldberg 1957). Vogeler (1981) traced the manner in which ownership and control were extended by agribusiness companies over an increasing part of all American farming. Some of the most capital intensive of farming practices, particularly lot fattening of cattle, and later the intensive production of pigs, fell largely into the hands of vertically integrated agribusiness companies that had interposed themselves between maize and soya farmers on the one hand and increasingly specialized livestock producers on the other. Except in the highly industrialized feedlot business, agribusiness concerns have been reluctant to enter direct production, and accept its risks. More commonly, where they have acquired land they have engaged managers, often former family farmers, who then have to produce what the company wants, and how they want it.

Events in and around the USA

It is easiest to follow events in and around the dominant power in world agricultural trade in this period, before turning to Europe, then bringing the two together again in the World Trade Organization (WTO) period. We begin with a 'crisis' in the early 1970s. A minor El Niño in 1965 had caused a mild flutter in the world wheat market, the enduring international effect of which was to accelerate international funding for the Green Revolution in Asia. A larger El Niño in 1972–73, the first major one since the early 1940s, led to more serious trading perturbations, characteristically exaggerated by nervous market behaviour, as well as by political coincidence. Grain stocks in the USA had risen very rapidly after 1965 and government was very keen to get rid of them. The USSR, in the midst of an attempt to raise living standards by diversifying food production, found itself short of grain. Dealing through the big companies, at American insistence, the Russians bought three-quarters of all the grain commercially traded in the world in 1972–73, leading to a major increase in prices even though sales to the USSR were substantially subsidized.

Developing countries now found themselves insufficiently funded to buy the imports to which they had grown accustomed. They were saved by the flood of dollars suddenly made available by the oil price spike of early 1973. They were able to borrow from banks that were very willing to lend these windfall funds. Other borrowers were American farmers, suddenly urged to plant from 'fencerow to fencerow', forgetting all about the conservation reserves of previous years. Soya was also involved in the boom, contributing in this case to sudden American export embargoes in fear of domestic feed shortages. Following a political outcry over the large subsidized sales of wheat to the USSR in 1972–73, wheat exports to the Russians were embargoed in 1974–75. All these embargoes were totally ineffective: the USSR got all the grain it needed from Canada, Argentina, Australia and other countries, and even indirectly from the USA via eastern European countries. This was handled by the large grain dealers, much of it not even becoming known to the US government (Tarrant 1980; Friedmann 1993).

The American grain trade with the USSR was put on long-term contracts in 1975, but the affair had lasting consequences. For importers, the dangers of reliance on a single dominant exporter became apparent, and led to the rapid expansion of EU (especially French) exports, to Japanese diversification of its food import sources, and in Brazil to expansion of maize and soya production and their manufacture into livestock feed.[3] Paradoxically, from the mid-1970s until 1991, food exports to the USSR became the lynchpin of the US agricultural regime, even as new markets were aggressively sought elsewhere. International grain prices began to decline in the mid-1970s as competition mounted and surpluses began to reappear.

Within the USA market support had been dropped in the early 1970s but, while market prices rose rapidly, many of those American farmers who had not previously invested in major production improvements or who wished to enlarge their farms did so with easy credit from petrodollars. This was a period of rising inflation, and there seemed no risk, as land values rose rapidly in the belief that investment in land is the best hedge against inflation. The amount of outstanding farm mortgage debt rose by 59 per cent between 1970 and 1980, creating a higher share of debt in farm investment than had been experienced since 1919 (Melichar 1977; Barnett 2003). Once prices fell again, the administration responded to pressure groups seeking restoration of support, but not sufficiently to shield farmers from their heavy debts.

The American farm crisis of the 1980s

World prices declined rapidly right through the early 1980s. National prices followed suit, interest rates rose sharply and many farmers again could not service their debts, creating conditions for a new farm liquidity crisis of dimensions potentially as severe as in the 1930s. In the USA 11 per cent of farms (220,000 farms) were lost in the recession of 1982–86, in fact far fewer than were less painfully lost in the prosperous 1950s and 1960s (Committee 1989: 54). Some areas suffered much more than others, depending on the debt vulnerability of farms and the viability in depression of the farm enterprises. For surviving US farmers the crisis quickly ended after the dramatically enlarged farm support programme of the 1985 Farm Security Act came into operation (Barnett 2003). Notwithstanding heavy debt loads, the larger and more productive farms had weathered the crisis much better than the smaller farms, which were a majority among those that went to the wall.[4]

Some marginal farms did survive, and from the 1970s onward two other substantial classes of smaller farms emerged, subsisting mainly on non-farm incomes that more than compensated for what were generally on-farm losses. These were farms operated by retired farmers with pension support, and the growing number of 'lifestyle' farms operated from the peri-urban and rural residences of families that depended on urban incomes. Collectively totalling more than a million farms accounting for 8 per cent of total sales in the USA in the late 1980s, these two groups received few support payments because of their generally small scale of operation. A proportion of the commercial farms were no larger, but operated to obtain maximum benefit from the system (Orden 2003, tables 1 and 2).

American farm policy in the 1980s

At the peak in 1981, the USA controlled 39 per cent of world agricultural trade. This share then declined sharply. The Farm Security Act of 1985 was the product of the crisis in progress while it was being debated, and also of the rise in environmental consciousness that had taken place since 1970 (Blaikie and Brookfield 1987). The system made use of all the tools developed since the 1930s, with the addition of a much enlarged soil conservation programme. With some temporary changes in the 1990s, the same formulae have continued into the present century. The main benefit was to producers of a set of major crops, now including maize, cotton, wheat, barley, sorghum, rice, tobacco, sugar and milk. To receive benefit, farmers needed to be enrolled in a commodity programme. For subsidy purposes, an average yield was applied to a base acreage which was the average of the past five years. Acreage could not be transferred to another crop without loss of benefit. Very clearly, these provisions militated against the conservationist use of crop rotations (Committee 1989). They endured for a decade.

The 1985 bill still relied on the old loan price, which the Commodity Credit Corporation (CCC) of the US Department of Agriculture advanced against a planted crop before harvest. This loan price in effect offered a guaranteed minimum price. If the farmer did not receive this price from the market, the crop could still be 'sold' in liquidation of the loan to the CCC until 1996, although from 1977 farmers had been subsidized to store their own grain on farm. The main difference after 1985 was that loan rates were substantially reduced, leading to a price drop that was translated internationally into a world price drop. At the same time, international attacks on production subsidies, exercised through negotiations under the General Agreement on Tariffs and Trade (GATT), had increasing effect.

Direct payments to farmers therefore assumed greater importance. The most important of these was now the 'deficiency payment', a payment depending on the size of the marketed crop and hence of main benefit to the larger producer. The 'deficiency payment' was the difference between the received market price and a 'target price' determined each year by the Department of Agriculture, with congressional approval. Target prices were below market prices only in exceptional years, and were generally set well above them.[5] To these measures were also added the Agricultural Credit Act of 1987, requiring the Farmers' Home Administration, successor to a previously weakly funded New Deal innovation, to make all feasible efforts to restructure loans including writing them off. It did so, to the tune of several billion dollars (Committee 1989).

One major element of the 1985 bill was to bring together conservationist measures that were coming to be seen as increasingly important. Support payments for acreage 'set aside' from production were first offered in the 1930s, and continued into the 1970s. The trade boom in the early 1970s had led to greatly increased production, however, with much of it on land sensitive to erosion. The 1985 Act consequently included a conservation reserve programme, as we will describe in Chapter 9. Large areas were to be set aside for long periods. All this new funding got most American farmers out of the hole they were in by the mid-1980s but it left a formidable legacy for the future.

The Common Agriculture Policy of the European Union

The idea of a European Union (EU) had been advanced by several writers and national leaders from the time of Napoleon through the first half of the twentieth century. It was taken up more seriously after World War II. After a preliminary period in which France and (western) Germany took the lead, the Treaty of Rome signed in 1957 formed an economic community of 6 countries that was progressively enlarged to 15 by 1995. A further enlargement to 25 countries took place in 2004 and then to 27 in 2007. The Common Agricultural Policy (CAP) was negotiated between 1957 and 1966, when all its provisions came into force. From the outset it has been an association of countries with a single external trading border. To a greater or lesser degree, all were subsidizing their agriculture at the time of entry. In the case of Britain, which finally joined only in 1973, doing so meant ending the preference given since 1931 to countries in the British Commonwealth.[6] Some of the former preferences that had been offered by Britain and other members to their former colonial territories, principally for tropical crops, were taken over in 1975 by the EU importing countries under the first of a succession of agreements signed in that year at Lomé in West Africa. This did not apply to temperate zone commodities that were also produced on a large scale in the EU member countries, such as wool, meat and dairy produce. The loss of preferential marketing in Britain most strongly affected Australia and especially New Zealand.

The European parliament has little say in matters of agriculture, which are delegated to the Commission and in practice to its agriculture directorate together with the council of national ministers of agriculture and trade. Every change involves compromise within a constantly evolving European Union, and small adjustments are easier to achieve than radical changes. There has therefore been nothing resembling the successive farm bills in the USA, and there was no major change in the rules of the CAP from 1966 until 1992. There were then further changes in 2000, and others, agreed in 2003 and 2004, are being implemented over a seven-year period from 2005. A large part of what was determined in 1958–66 remained in force until these latest changes.

The nature of the Common Agricultural Policy

As the EU is an area of internal free trade, with a single external tariff, all the input and production subsidies that were applied in individual countries had to be replaced by a unified set of policies. Agriculture was for a long time the only area covered by a common European policy, and is still the principal one, absorbing more than the structural programme. The amount of the collective national GDP devoted to the EU budget remains only a small fraction, around 1 per cent, but has grown absolutely over time. At the end of the 1950s, European agriculture had recovered from the dislocations of World War II, and had made great progress in enhancing productivity (Lamartine-Yates 1960). But although continental self-sufficiency in all but tropical crops was clearly attainable, a large dependence on overseas imports had already re-emerged in most countries, and the first goal of the CAP was to increase production. In the still uncertain political climate of the 'cold war' period, and with memories of the huge unemployment levels of the 1930s still very much alive, this was coupled with an intention to keep farming as an attractive occupation in relation to industrial and other work. Initially, this was to be paid for by the consumer rather than by the taxpayer, through higher over-the-counter prices generated first by high external tariffs and second by subsidized farm-gate prices. Where individual countries had, before entry, imposed lower tariffs than the EU standard, the costs to the consumer were not insignificant.

For some time the only measure used was the guaranteed annually determined farm-gate price, apart from uniform external tariffs which came into effect over a period of years. It was not unlike the American loan price in many respects. When world prices fell close to the guaranteed price, the participating national authorities would buy surplus production at that price and subsidized wholesalers would store it. The high tariff offers substantial protection, and when associated with an accurately estimated world price, the system worked. But for much agricultural produce the amount traded internationally is only a small part of world production, so that the 'world price' is often very hard to determine. For wheat and other grains the Chicago grain exchange price is commonly used.[7]

National marketing authorities, and later the farmers and trading companies themselves, received storage subsidies. After a short time they received further subsidies to export the surplus production elsewhere in the world. This applied to grain, oilseeds, beef, tobacco and milk (in powdered form). With only minor exceptions, support was not limited by quantity. The effect of the subsidies was that internal demand was so soon satisfied and during the 1970s the EU became a major net exporter, though more of food products than of the basic crops themselves. The system of subsidies encouraged larger farmers to expand production massively, and they were more able to do so because the constant flow of new technical advances in both arable and livestock farming lowered the cost of production.

The CAP proved increasingly costly, especially during the years of turbulent primary produce prices after the mid-1970s. As production continued to soar, the budget was seriously threatened with exhaustion. New essays with 'set-aside' land failed significantly to check expansion, and in the late 1980s outlays had to be capped. Automatic price cuts were introduced when production approached maximum quantities that could be stored. This was the time of the infamous 'mountains' of wheat and beef and 'lakes' of milk and wine purchased at intervention prices. Milk production, at least, was checked by the introduction of quotas which could be traded, extending down to the level of the individual farm. In a short time, the quotas became valuable assets. In 1992, the first major reform of the CAP was agreed, setting out to partially replace market support prices with direct payments to farmers. The necessity for the 1992 reforms lay principally in the financial crisis experienced by the CAP in the 1980s, together with the prospect of an EU expansion after the collapse of the Russian-led system in eastern Europe. It did not arise specifically from the rising storm of criticism of European and American subsidies from trade partners, but the reforms did facilitate the international trade negotiations then in progress. Support prices for grains were cut by 35 per cent over three years, making EU cereals more attractive to the animal feed industry and thus replacing imports of American and other maize. This breach of the 1950s agreement with USA almost triggered a trade war!

A consistently stated aim of the CAP has been to generate an adequate income for as many as possible of Europe's farmers, but this has never been the outcome in practice. As in North America, most of the financial benefits have gone to the larger farmers and landowners, including some very large landowners, such as leading members of the British royal family. In 2002, when the EU first released data on its payments, 70 per cent of farmers received less than 5,000 euros in subsidy, and more than half the direct payments went to a minority of individuals who had incomes exceeding that of the average taxpayer. As Bureau (2003: 54) remarked 'The CAP therefore organizes wealth transfers to a social group with incomes that exceed the average wage of those who fund this transfer.' All this is much the same as the outcome in the USA. For the rest of the rural population, off-farm employment became even more a major source of income than in North America. By 1997, four out of five people involved in agriculture worked only part-time on farms, and only half to two-thirds of the total farm income came from farming (Bureau 2003). In Greece, where most farms remain very small, part-time working in agriculture characterized 90 per cent of the farm population.[8]

The North Atlantic and the world: the World Trade Organization

The crisis of the 1970s exposed the extent to which international commerce and finance had escaped the control of governments. A market economy dominated by corporate power now made its formal appearance in the ideological costume of neo-liberalism (see Chapter 8). A principal role of state management quickly came to be regulation of the social and economic order so as to permit free operation of this new sort of market economy. This did not happen easily, as the system had generated powerful vested interests. Even the facts of what was happening were in dispute. One hotly contested area over which many half-truths flourished was the comparative subsidization of international trade in food and feedstuffs. For example, data were produced in 1988 indicating that American support, as a percentage of producers' income, was below levels in Europe and Canada (cited in Committee 1989: 97). But this has to be viewed in quantitative perspective. The whole Canadian wheat area was only about the size of the additional wheat acreage planted in the USA in the 1970s alone. Canada was trying to adopt a hands-off policy at the time, and in any case lacked the resources to support its farmers on the American scale. Europe was another matter, however, and by the 1980s it had become a very real competitor to the USA in world agricultural trade. The 1985 US farm bill responded to this competition and was, in the view of McMichael (1994: 644) 'aimed at restructuring the world food market by drastically cheapening the price of US agro-exports', that is to restore the situation to what it was before the mid-1970s.

In this highly competitive situation it took a full decade to negotiate the 1995 world trade agreement which established the World Trade Organization (WTO). It set up rules with which signatory countries were required to comply, dividing supports to agriculture into three 'boxes'. The most important was the 'amber box' of subsidies and other measures that distort trade. A 'green box' included measures not held to be trade distorting, and therefore not subject to any restrictions under the agreement. Under pressure from the Europeans as well as the Americans, an intermediate 'blue box' was created to cover otherwise 'amber' measures which applied only to partial acreage, including therefore all those associated with land idling (Bureau 2003: 64). The agreement also required substantial tariff reductions over a ten-year period. It made only limited progress toward the goal of anything that could properly be termed 'free trade', but it entrenched conditions that would favour the most powerful trading countries, and the largest trading interests within them.

Reforms, and a shift in emphasis

The EU-CAP reforms of 1992, discussed above, entailed adaptation to this emerging structure, with the purpose of influencing its shape. In 1996 the USA undertook an important unilateral reform, with the farm bill of that year. It made major changes, eliminating 'target prices' and, importantly, wiping out the 1985 restrictions on changing land use on the registered 'base acreage'. Instead, direct payments were to be made on the basis of land use allocations and average yields over previous periods in the 1980s, whether or not the land was in the same use. These payments were independent of current market price. Land idling requirements were also relaxed. The basis of these direct payments was 85 per cent of the base acreage, thus falling into the compromise 'blue box' of the WTO rules and escaping restriction. By removing these restrictions, the 1996 bill offered 'freedom to farm' which was soon taken up. Loan payments continued, but these were now cash payments, and farmers no longer forfeited crops into government storage. Instead, both national and foreign marketing were now more completely taken up by agribusiness.

In Europe, the new area-based payments were 'compensation' for loss of market support, conditional on all but the smaller landholders setting aside from 5 to 15 per cent of their crop land in each year. Payments were determined on the basis of regional yields for periods in the 1980s, and thus differed substantially between both countries and regions within them. The direct payments, which were more widely used to replace intervention buying from 2000 onward, nonetheless remained coupled to the area under cultivation in the given year. In 2003 and 2004 a new set of reforms was agreed, coming into effect in 2005. Direct payments were uncoupled from current production, as in the post-1996 American system, creating a 'single farm payment' the value of which is based on subsidy receipts in 2000–02, and for which the only condition is that the farm be kept in environmentally acceptable condition. Actual sale prices are therefore those of the open market, to the distress of many farmers who do not see the direct payments as effective compensation. They are, in any event, probably rightly regarded as politically insecure. The 2005 changes are of a far-reaching nature and reflect new internal and external conditions. Since there are options to continue some forms of linked payment, and these have been taken up by some countries, it will not be until after 2008 that the 'single farm payment' applies to the whole EU.

In Europe, to a much greater degree than in the USA, the new provisions have been intended to serve environmental purposes, as well as to check 'surplus' production. The effect of modern agricultural practices on the environment has been very much the same as in the USA, with the quality of drinking water affected over large areas, and hedges and fences removed in the interests of using large machinery. Starting in a small way in the 1990s, therefore, and much more seriously since 2000, environmental management has been built into the CAP. From 2000, 'eco-conditionality', following practices that are environmentally friendly, has become a qualification for receipt of direct payments. The CAP now officially has a 'second pillar', consisting of a set of environmental and regional measures offering the largely urban taxpayers, who like to visit and enjoy the country, more for their euro while hopefully also revitalizing depressed rural areas. Some earlier attempts to achieve this latter end, in Italy, are described in Box 6.2. The new environmental issues are discussed in Chapters 9 and 10.

American policy continued to be more volatile. The farm bill of 1996 did not endure long. While 1996 was a year of high world prices for agricultural goods, these declined over the next several years, and the organized farmers soon applied pressure for new subsidy measures.[9] They got them principally in the form of 'market loss assistance' payments, which by 2001 were providing over 40 per cent of direct farm income. All these interim steps cost a lot more than had been budgeted in 1996, and paved the way for the farm bill of 2002. Target prices were fully restored in this bill, and authorized for a number of other products, especially oilseeds, that had not been covered in previous years. Loan rates, which were also extended to a range of new products, were now fixed in nominal terms. Farmers were given the opportunity to declare new base acreages and yields, although the degree to which yields could be varied differed between commodities. Ability to change land use remained untouched from 1996.

The total cost of support was increased by the 2002 bill, but largely in ways that were not tied to production, thus reducing the likelihood of exceeding the negotiated WTO limit on 'amber-box' subsidies. While the 2002 bill has been hotly attacked

---

**BOX 6.2 EARLY REGIONAL DEVELOPMENT IN
EUROPE: THE ITALIAN MEZZOGIORNO**

Until the 1990s, the CAP had little direct involvement in regional integration and harmonization. The EU, as a whole, offered subsidies for development of the 'lagging' areas from 'structural funds'. But there have been important government programmes at national level since before World War II and these have continued and been intensified, especially in Italy. Franklin (1969) provides what is probably the best survey in English of progress up to the early EU years. The Italian land reform of the post-war period was accompanied by a major drive to develop commercial agriculture. There were some contradictions. Commercialization favoured the larger landholder or tenant whether or not employing labour; on the other hand many of the small farms created or enlarged by the redistribution still had insufficient land for profitable commercial agriculture, even with capital investment assisted by the state. Under continuing plans involving irrigation and other forms of infrastructure investment in most southern regions with suitable topography, it was hoped that the Mezzogiorno would develop into the premier region within the EU for production of Mediterranean crops for the multinational market. After 1965, planning placed greater emphasis on industry and especially heavy industry, with mixed long-term results. Farmers were encouraged to retire early, with a good pension scheme, in order to free up land for more 'efficient' enterprises.

Already in the mid-1960s, it could be seen that what was happening was a great improvement in regional income, but without self-sustaining development. 'Instead, it has been a massive welfare dependent programme of support and redistribution, engendered and maintained by the expansion of wealth in the north of Italy and in other member countries of the [EU], together with considerable aid from the U.S.A.' (Franklin 1969: 174). Ten years on, Wade (1979) reached a similar conclusion. The heavy dependence of the region on the state, meaning in particular transfers from the north Italian taxpayer, remained a keen issue until the present day. One family in five still lives in a situation of 'economic distress'. It is now argued that agriculture – which received all emphasis at the beginning – has been neglected in favour of the public sector and retailing (Villers 1998).

---

by critics at home and abroad, especially in countries that compete for exports with the American food and fibre sector, it could be claimed on its behalf that it is merely doing what other countries are doing in finding 'clever tricks that can be used to meet the letter of their WTO commitments without substantially changing the support provided to agriculture' (Orden 2003: 30). The 2002 bill has cost more than its predecessors, and offers yet more ample levels of support to the large producers and associated agribusiness enterprises. The benefit limits to individual producers are more generous than ever before, and the same means exist to get around these limitations.

Prospects for change

As we write, serious negotiation on a new US farm bill is in progress. In Europe there is to be a review of the progress of the CAP reforms in 2008. Strong views regarding an end not only to production subsidies but also to direct payments to farmers, except for environmental services, have been expressed, especially by the UK government. CAP expenditure as a whole has been capped until 2013. Production and especially export subsidies have been under major attack in the WTO. In 2001 the Doha round of WTO negotiations began with the object of further opening markets in agricultural products, manufactures and also services. It quickly ran into trouble as, for the first time, a strong block of developing countries led by India, China and Brazil fiercely attacked the North Atlantic farm and export subsidies, declining to open their own markets without major concessions. Through a series of stormy meetings, beset by small armies of unofficial protestors outside in the streets, they got as far as gaining a promise to eliminate export subsidies by 2013. Incompatible proposals were put forward by Europe and the USA, but by 2006 the meetings failed to reach any comprehensive new agreement. Although attempts are being made to revive the talks in 2007, everyone seeks gains and is indisposed to concede anything without such gains. The fact that there is a new politics of world trade, in which the North Atlantic countries find it hard to dominate as before, has come sharply into focus during this contentious process. The debate of export subsidization is central, and is developed in Box 6.3.

Yet a system that grew up in the 1930s has remained resilient in two continents, and up to 2007 was still expanding, territorially in one continent and financially in the other. The system grew up to protect the farmer, but as production has expanded with the aid of this protection, so also have the upstream and downstream sectors of the food and fibre industry, growing to be very much larger than the farm sector itself. It is also significant that the rural vote is of major political importance especially in countries where there is a fine balance between political parties, which is true on both sides of the Atlantic. Farmers themselves may be only a fairly small part of the whole rural vote, but many of their neighbours derive their incomes from other parts of the food and fibre industry, and identify with their interests. Many of the firms in this large industry have also benefited from the subsidies, both indirectly and even directly. While public discussion focuses on the farmer almost as exclusively as it did in the 1930s, very large, if unrecognized, industries are now involved.

No one is really satisfied by present developments. When the USA introduced direct payments to its cotton growers, Brazil, supported by a group of other countries, complained to the WTO that these were still production subsidies that distorted the market. In 2005 they won their case. The European 'single farm payment' is more specifically divorced from actual land use than the American payment, but there

---

BOX 6.3 THE DEBATE OVER EXPORT SUBSIDIES

Europe's export subsidies have been direct and very transparent. They have therefore received a high proportion of the obloquy that fell on the exporting nations. By late in the twentieth century, Europe was by far the leading directly subsidizing exporter, supplying some 80 per cent of all direct export subsidies paid within the countries adhering to the WTO. But the USA and other countries have encouraged exports in different ways, especially by means of 'export credits', loans from national institutions to overseas companies or government agencies to finance the purchase of goods, then repayable at commercial interest, or else to accumulate as state debt. They also provide insurance to the exporter against losses due to non-payment, and sometimes guarantee commercial bank loans. In the early twenty-first century it was claimed in defence of European practices that 84 per cent of export credits in world agricultural trade was provided by the USA, Canada and Australia (Bureau 2003: 82). Canada and Australia insist that their farmers are not subsidized. However, the total range of measures short of subsidy includes not only export credits and insurance programmes, but also concessions of various kinds and allowances made in national tax systems. Without a need to express disbelief, it can be said that selective use of statistics is capable of 'proving' a great many things. In addition to official statements by the countries themselves, there are many Internet sources on this topic, some of them pointing out that the individual EU countries also provide export credits. They are used for industrial exports more substantially than agricultural exports, and hence Japan is probably the largest provider of this form of finance. There are substantial discrepancies between different sources on the degree to which these credits are manipulated to become de-facto subsidies, rather than merely providing back-up support and insurance for private-to-private transactions.

---

have already been many complaints that this is just another way of encouraging farmers to produce more, and thus is still 'trade distorting'. Trade partners and competitors, and the farmers themselves, are not happy with the new arrangements. Farmers trust supported prices but fear their direct payments are vulnerable. They have reason: in Europe and in other developed countries there is growing dissatisfaction at the transfer of public funds to the small farming population. The whole system runs against the free-market ideology that prevails in both world regions, and it certainly fails to achieve redistributive justice. Yet there are farmers in the developed world who have had to manage without these supports, and have succeeded in doing so despite losses and severe hardship when such support as they once had has been withdrawn. We meet some in Chapter 8.

Notes

1   The European Union, as such, has been in existence only since ratification of the 1989 Single European (Maastricht) Treaty. In earlier years it had a variety of names. For simplicity, we use European Union (or EU) throughout.

2   The Canadians were leaders in the breeding of oil-seed rape (*Brassica napus*), hence the naming of many new varieties as canola. Since the 1980s, the increasing trend toward political correctness in terminology has led to the much more general replacement of the term 'rape' by 'canola' in anglophone public use. A much older alternative term is 'colza'.

3   This 'value-added' approach also applied to ethanol made from Brazilian sugar, in response to the hike in oil prices. Both these initiatives were taken under the military regime that ruled Brazil from 1964 until 1983.

4   Losses were proportionately more severe in Canada where the state was unable to ease the decline in farm prices significantly – and did not wish to. Neo-liberal free-market theories were already coming into play, and the Canadian objective in this period was to sustain the national economy in the context of what was already being termed 'globalization'. This involved a preference for the 'more efficient' larger farmers who could better manage by themselves. In the agriculturally marginal areas of Canada there was a sustained abandonment of farmland during the whole period after 1950.

5   There were limits to the payment per farm, which many large and corporate farmers avoided in such ways as notionally subdividing their properties. Most subsidies have always contained restrictions on the amount of benefit that might go to any one farm. The manner in which larger farms were able to avoid losing government subsidies from these measures, by subdividing large farms into smaller units still under collective or corporate control, is described at length by Vogeler (1981: 164–70). His discussion includes a verbatim reprint of a revealing set of exchanges between two leading members of Congress before a Senate Committee inquiry in 1971–72.

6   It also meant giving up a system of deficiency payments which supported farmers directly without influencing prices.

7   In 2000 and 2001, excellent harvests in Russia and Ukraine brought the 'Black Sea' prices so far below the Chicago price that this source of wheat could profitably be imported despite the tariff. For two years, Europe became the world's largest wheat importer.

8   In the USA, using 1988 data, farm income was 27 per cent of total income in a representative set of regions. The proportion varied from 47 per cent in the western corn belt and northern plains to 15 per cent in the 'eastern highlands', meaning the Appalachians (Committee 1989).

9   Mainstream farmers' organizations had already become politically powerful in some countries as early as the 1940s, when the British National Farmers' Union had a big say in the framing of post-war agricultural policy and even in determining the levels of subsidy. They became steadily more powerful during the subsequent decades and retain considerable influence over policy today, notwithstanding the declining economic significance of farming within national economies. The mainstream organizations have not been effective representatives of the smaller family farmers and since the 1980s a number of breakaway bodies has emerged. We discuss them in Chapter 14.

7 Farms collectivized and de-collectivized

Russia and China

While farms in the North Atlantic countries were experiencing government support on an unprecedented scale in the years after 1930, a different sort of state intervention was being pioneered in other lands. The purpose was the creation of socialist societies. Collectivization of farms was a major innovation of the twentieth century, and at its peak in the 1970s it was applied in a large part of eastern Europe and in the socialist countries of Asia. A number of developing country governments in Africa and Latin America also set up cooperative or collective rural institutions in the post-1950 period, but none endured except in Cuba. From the 1980s onward, the trend was reversed and collective institutions were dismantled in country after country. Our discussion is limited to Russia and China. In these two countries collectivization was most thoroughly applied. We write mainly about China, but first briefly introduce the Russian experience.

Russia

Between 1928 and the early 1930s the farms and villages of Russia became a set of collective and state farms. Collectivization had both economic and political objectives, and the economic one was only slowly achieved. The political objective, that of destroying a rich farmer (*kulak*) class thought to have grown in size during the 1920s after the 1917–19 redistribution of the large estates, received the first and brutal priority. The early collectives were of village size, and after regrouping land into larger blocks, they were initially worked in much the same way as the individual farms. Progressively through the 1930s, mechanization was introduced through a network of 'machine and tractor stations', each of which served a group of collectives, together with the labour-employing state farms that were also set up. Collectivized farmers retained their own houses, small livestock and usually a family plot. The family plots became a legal entitlement in 1935, with the proviso that they not exceed 7 per cent of the whole collective farm area. Women did most of the work on them. In effect they were small private family farms, with the difference that while the produce remained the family's own for use or sale, most material inputs came from the collective land and livestock. Often the family plots received as many hours of family work as the collective fields. Toward the end of the Soviet period, when controls on their use were relaxed, the intensively managed household plots seem to have developed something of the classic pattern of infields in relation to the collective outfields beyond them.

In Moscow Oblast as a whole, there were 6,000 collective farms in 1940, occupying 85 per cent of the agricultural area (Bayliss-Smith 1982). After World War II, during which all the western half of Moscow Oblast had been fought over, including Volokolamsk, described as it was in 1910 in Chapter 2, the whole farming system had to be re-established urgently to meet the pressing demand for food. Many of the collectives had ceased to operate effectively because of heavy loss of life during the war. Many of them were replaced by state farms, using only employed labour. The remaining collectives were combined into larger units. For a time, there was a move to base them on the machine and tractor stations, which would have made them huge, but instead these stations were broken up in the late 1950s and their equipment dispersed to the collective and state farms. By the 1970s in the Moscow region, there was heavy emphasis on animals, both for meat and milk. Private plots supplied the city vegetable markets (Bayliss-Smith 1982).

Changes since 1991

Contrary to the expectation of many observers, land reform in 1992–94 did not lead to the rapid break up of either collective or state farms. Farmers and workers acquired shares, but they were not tied to specific areas of land. Unlike some of the east European countries in which collectivization did not begin until after 1945, no attempt was made to restore farms within their old boundaries. In all Russia, only about 300,000 households used their right to withdraw and set up their own farms. Deprived of the collective inputs and access to sufficient credit, many of these failed. Others survived successfully, earning incomes above those of the managers of state and collective farms. Wegren *et al.* (2003) suggest that this has been due more to entry into small-scale agribusiness than to enhanced production.

With the aid of political support at the regional level, most collective farms in Russia managed to survive the efforts of the Yeltsin government to destroy them in 1991–94. During the Soviet period they had provided valuable fringe benefits to the farmers, including free education and health care, tax concessions and paid vacations. Since 1992, these advantages were severely eroded. Yet in Russia as a whole 86 per cent of farmland was still run by large enterprises in the year 2000 and 6 per cent was in household plots, with only 7.9 per cent leased and operating privately outside the system (Serova 2002). One study in Moscow Oblast found a mixed landscape of private and collective farming, in which patron–client relationships between the managers of remaining collectives and the now semi-independent farmers continued to form part of the new social system (Pallot and Nefedova 2003). Since 2000, it would seem that more stable economic conditions have been of advantage to all.

Chinese farmers and the party: 1949–78

The transition from the old rural China (Chapter 2) to the 'great leap forward' of 1958–59 took only a few years, but it was not as immediate as the Russian collectivization, and went through a series of quite well-defined stages. It began in the early and mid-1940s in the areas that were already controlled by the Peoples' Liberation Army, and then spread to the rest of the country as the army advanced. Following rent reduction and a fiercely progressive system of taxation, land reform began in 1947 in areas already under communist control. It was enacted nationally in 1950, and completed throughout the country by 1952. All households were classified into groups that would lose or groups that would receive. Landlords and some 'rich peasants' who lived principally on rents and the labour of hired workers, lost all or most of their land and productive assets. A significant number, from 700,000 to more than a million, were killed (Kerkvliet and Selden 1998). Most lineage and other group property was expropriated. The poor, those holding either insufficient land or none, got what was redistributed. The middle-ranking group of farmers was not targeted for expropriation or redistribution. While better planned than the Russian expropriations of 1918–19 and 1928–32, this reform still did not create a fully egalitarian rural society.

It was intended that land reform should be followed by a transition to socialism through cooperative work. This stage drew heavily on mutual aid arrangements that had been common for centuries, but it set out to organize mutual aid among defined teams, with leaders, and operating around the year so that off-season work and 'sidelines' could be brought within its ambit. In some areas, mutual aid teams began even before the land reform was complete. By absorbing the labour of the poorer farmers, they reduced the amount available for hire by those who were better off. Grain trading became a state monopoly in 1953, eliminating opportunities for those with more resources to profit from trade, and also to secure state control over supplies. The more entrepreneurial among the better-off farmers were also discouraged from money lending by the credit unions that were set up in most of the administrative villages (Shue 1980).[1] From 1953, mutual aid gave way to 'voluntary' cooperative associations which pooled all productive resources among private farmers, and then in 1956–57 to less voluntary 'higher-level' cooperatives that also pooled all land and labour. In these 'socialist' cooperatives, no private property remained.

There are several good accounts of these events, including Chan *et al.* (1984) and Siu (1989). Hinton (1983) provided a lengthy discussion in which he reproduced the basic constitution for cooperatives that was finally promulgated in 1956. He also reconstructed in imaginative detail the course of events in the Shanxi near-urban village he called 'Long Bow' where he had witnessed the land reform in 1948 (Hinton 1966).[2] A more intimate account of events during the 'mutual aid team' stage, in another part of Shanxi, is given in a 'novel from life' by Liu Ching (1964). It is built around a young farmer, son of a poor tenant farmer, who became a team leader, succeeded in all his enterprises and became first chairman of the cooperative when it was set up.[3] The hopes and tensions of this period are better captured in the novel than in most historical writing.

In 1955, a system of inheritable residence permits (*hukou*) was set up, dividing the national population into two main classes, urban and rural, based on residence, and only to some degree on occupation. Rural people were specifically tied to their villages. This kept most farmers on the land until the laws began to be relaxed in the late 1980s. Until after 1978, all those who were classed as rural residents, except for those with a bad class background, were regarded as naturally revolutionary, while the non-peasants, who had ration books and other privileges, paid for their advantages by being suspect of bourgeois tendencies. Yet the urban *hukou* offered higher status, and could be achieved by peasants only through high educational achievement, army service or sponsorship by a higher-level cadre (Kipnis 1995). Through these measures, the state, meaning the party centre, took an increasingly directive role in the affairs of the nation, including the supervision of both rural and urban cooperatives and their memberships. By the mid-1950s, local administrations and collectives were already receiving ambitious targets for produce delivery from the central authorities. By this time, some of the new rural cooperatives drew all farmers into their membership because the better-off could no longer find enough labourers, make incomes from renting land or make money in trade or by lending. Other cooperatives were less successful, and a significant number were wound up by central direction. This was not only the pursuit of efficiency; as in Russia in the 1920s, there developed some sharp differences about the speed and direction of rural change among party leaders. The goals of collectivization included grasping economies of scale with the aim of mechanization and the mass mobilization of rural labour. Behind this was the overriding aim of extracting the rural surplus to fund China's rapid industrialization. It was achieved by taxation, compulsory procurement of produce at low prices and pricing of inputs so as to worsen the terms of trade for the peasants (Ash 2006; Kueh 2006). Although the methods were distinctive, it was all very much in line with contemporary worldwide development thinking.

Collectivization increased in 1955–56, then in 1958 with the 'great leap forward' the collectives were grouped into very large communes, some of them covering whole rural counties, with central direction of all aspects of production and distribution. Consumption was also centralized at the brigade or team levels by the provision of meals from central kitchens. Sidelines, where not discontinued during the 'leap', were incorporated into the commune programme. The pre-revolutionary pattern of rural industry was brought into this system. It mainly supplied agricultural inputs or processed farm outputs, but was famously enlarged to include even iron and steel production at local level. The objective was rural self-sufficiency. Central direction, through the province and commune, was carried down to the team, household and individual, with the consequence of total disruption of agricultural work schedules and farm planning.

This was an organizational change of huge dimensions. Remote direction of farming activity, and of the allocation of labour between farming, public works and the sideline industries, succeeded in only a minority of communes. The pressure on cadres to report dramatic successes, whatever the reality, destroyed the statistical base that might have supported better planning (Shue 1980; Ash 2006). Thus the good yields of 1958 were projected as even higher reported yields in 1959, when production was in fact far smaller. These failures, which encouraged very heavy exactions of grain to supply urban workers and even export, together with the mobilization of large numbers of workers for tasks at a distance from home, contributed substantially to the widespread crop losses and famine of 1959–61. The response was a substantial, but only partial, retreat. The huge communes were broken up into smaller units, but all smaller-scale collectives remained. Within only a year or two more, central direction, fierce egalitarianism and the mobilization of peasants for public works were resumed in the 'great proletarian cultural revolution' which began in 1966. In the early years of this ten-year movement, both private plots and sideline activities were decried as 'capitalist'. In some places and at times the sideline activities were again banned, so that all effort could be devoted to the priority aim of 'grain first'. The private plots, always subject to reallocation or confiscation through the collective period, were also sometimes and in some areas eliminated (Oi 1989).

Chinese farming in the collective period

The fully elaborated collective system in China lasted for little more than two decades, but it has had lasting effects. Writing at the very end of the collective period, Huang wrote that:

> The removal of the supra-village gentry and the village notables of old has brought state power much more deeply into the village, buttressed by the apparatus of the party. At the same time, the scope of the state's power over the village, and of the village collectivity's power over the individual peasant, has expanded enormously through collectivization of production, the imposition of a planned economy, and the politicization of the population.
>
> (Huang 1985: 308)

The basic unit in the Chinese collective system was, as in Russia, the village (*cun*). It was usually the village population that constituted the brigade (*dadui*), the effective unit of management within the larger communes. Production teams were commonly formed from a whole small *cun*, or a section (*zu*) of a large village, sometimes dominated by one lineage. Whereas the commune brought together people who otherwise had little interaction, the brigade and production team were made up of people who knew one another well. In common with the family farm, they were units of both production and consumption, and similarly, they could not fire any of their membership. The average size of a commune in the early 1970s was about 3,000–3,500 households, that of a brigade about 200–250 households and of a production team about 30–40 households (Oi 1989: 5).

To farmers and their families in the collectives, the team was the central unit of activity and the basis of livelihood. Teams were run by committees of three to five, but the principal functionary was the team leader who determined what each family did and how they fared:

> Each morning the team leader blew his whistle to signal the start of the workday; team members met under the village tree or at the village gate to be told their day's work assignment or to consult the assignment sheet posted by the team leader.
>
> (Oi 1989: 137)

He also allocated means to earn work points, of value for extra grain rations but also for a share in any cash from the net income of the team. In the whole period down to 1978 these cash allocations were small, and often there were none. Inevitably, patronage developed under this system under which access to private plots was also determined. Team leaders were often, though not always, members of the Communist Party; leaders of brigades and communes always were. These cadres (*ganbu*) occupied the ambiguous and sometimes difficult position of being inter-mediaries between the party and state, the local authorities and the farmers. Under 'unified procurement' team leaders had to negotiate how much grain their members would deliver for compulsory sale, in competition with other teams; decisions would be discussed in the brigade and passed up to the commune. The amount for a commune had to match the requirement passed down from the centre through the province and county (Oi 1989). While the cadres at county and provincial levels had civil service employment conditions, the team, brigade and commune cadres reverted to the status of ordinary citizens when no longer fulfilling an official role. They were therefore more closely bound to village interests. They passed on the production targets and had to ensure that the required quota deliveries were actually produced. The quotas were determined as grain crop, or specialized cash crop promoted in particular areas. The purpose of the system was to keep the wheels of an industrializing economy oiled and running smoothly, by extracting the state's share of the 'surplus'. Although sales back to farmers moderated the effect on rural consumption in bad years, there was in fact no significant improvement in rural diets in the whole period from 1949 to 1976 (Ash 2006). Kerkvliet and Selden wrote of this period that:

> Bound to land they did not own and could not leave, villagers were locked into a system that denied them control over their labour power and the surpluses they produced, prevented them from selling or buying much in the shrivelled markets, and limited them to incomes in kind pegged at a low subsistence level.
>
> (Kerkvliet and Selden 1998: 48)

This is not far from saying that the ordinary members of the collectives had been reduced close to proletarian status, receiving little benefit from the collective gains. Such benefits were largely appropriated to sustain national programmes.

Strong team and brigade leaders who could command the loyalty of their members could sometimes achieve significant successes in their confrontations with higher-level cadres. Hinton (1983) recounts the savage infighting that could occur at local level, including conflicts between different production teams that supported one or other faction in the sometimes violent disputes of the Cultural

Revolution period. After about 1972, when the pressures eased, team and brigade activities became more autonomous, and sideline activities of all kinds, banned in the 1966–70 period, began to blossom as businesses. Most of the larger ones became the 'township and village enterprises' (TVEs) of the succeeding period. After 1976 they began to receive growing support from the authorities.

One aspect that the collectives did not quickly develop was mechanization, and it became a major force only toward the end of the collective period. China began to manufacture tractors only in 1958. In the party centre as well as among the farmers there was fear that the rural economy could not stand the major loss of employment from mechanization. It was felt that it would entail a return to the old underemployment of rural labour that collectivization had largely ended by setting up works programmes to occupy slack periods. Thus much work in the fields continued to be done by hand with oxen or other livestock, and only a modest number of tractors was introduced, principally in the wheat growing regions of the north. Hinton (1990), who had become a specialist in the mechanization of agriculture while farming in Pennsylvania between periods in China, was appalled that the major Chinese search for new industrial technologies in the 1980s did not include new technologies for agriculture. Although hand tractors, and low-powered tractor-like vehicles mainly used for transport are now very widespread, a large degree of reliance on manual labour in field operations has continued in the rice growing regions. Even by the time of the first agricultural census in 1997, there were still only 58 machines of all kinds per 1,000 persons engaged in agriculture (Fanfani and Brasili 2003). As late as 2000, at least in southern China, tractors were to be seen on the roads more than in the fields.

The economies of scale that were captured in the early days were fairly soon exhausted, and by 1970 agricultural production was stagnating. Substantial production improvements came as a Green Revolution programme was implemented through the collectives, but with low and fixed prices these improvements were not translated into higher income and better welfare for the farmers, except in so far as they were permitted to diversify their production. Discontent led to rising pressure from farmers and local leaders to expand the scope of household production and the market (Zhou 1996). After 1976, the party centre also took on board the need to improve incentives for higher or more diversified production, leading to a major improvement in procurement prices in 1978, and the initiation between 1978 and 1985 of other changes that broke up the collective system of production.

Rural China after 1978

The household responsibility contract system

Local experiments in contracting responsibility for tracts of land to individual farmers were briefly undertaken in the aftermath of the failure of the 'great leap forward', and were resumed in the 1970s at the same time that a number of other measures designed to raise the level of incentives were instituted (Bramall 2004). Such measures were supported for wider adoption by the party central committee at its 1978 meeting. Thus, partly from above and partly from below, the 'household responsibility contract system' was born. It came into general use in the early 1980s, in effect restoring family farming. Initially signed only for short periods, contracts were later extended for first 15 and then 30 years, and became both renewable and inheritable. The system extended to almost all of China between 1981 and 1985. At the same time the class classification of the population, until then inherited from the land reform period, was abolished but the urban/rural *hukou* system was untouched.

Land remained the property of the collectives, and still does, but was worked by individual households. Land shares were allocated to households on the basis of their size. This 'second land reform' was therefore more egalitarian than the 1950s land reform, which had left the land of 'middle peasants' intact (Shue 1980). It was supposed to last 15 years, but as household numbers changed through time inequalities quickly re-emerged. In some collectives there were further redistributions at intervals to moderate these trends, and in the process holders of minority surnames in at least some southern villages dominated by large patrilineages found themselves allocated only inferior land, deprived of irrigation water and discriminated against in other ways. They were thus forced to migrate to their 'ancestral' villages where, at least in principle, they should not experience discrimination (Liu and Murphy 2006). Before 1990 it generally became legal to rent out land, and once again also to hire labour. The terms 'commune', 'brigade' and 'team' were dropped, and the old hierarchy of local government – prefecture, county, township (*xiang*) and *cun* – was restored. The *xiang* became the lowest level of administration, and the *cun*-level work teams were abolished. With this also came decentralization of management, quota and tax collection. Tax collection was individualized, with consequences discussed under the 'peasant burden' in the following section.

While de-collectivization was generally welcomed, it was resisted in communities where the collective system had been more successful. Once de-facto privatization became general policy, stubborn collectives were very heavily leaned on by senior cadres from the township and county levels, but even so a minority was able to resist. After grain production attained a record level in 1984, and then sagged, marketing conditions were eased in 1985. The briefly eliminated grain delivery quotas were restored under different names, but the markedly lower price received by the farmers for the quota grain was eased in stages, so that by the mid-1990s only the negotiable market price applied. At the same time the provision of grain rations to the urban population ended. The state monopoly on grain purchase was removed, but since it remained technically illegal for private buyers to acquire grain directly from farmers, the provincial Grain Bureaus continued to buy the bulk of the crop. The price was varied according to the conditions of supply and demand, and if farmers were not satisfied with the price they could, if they were able to do so, withhold supply beyond the tax requirement (Watson and Findlay 1999). There was a smoother increase in the production of vegetables, fruit and other non-staples to which quotas did not apply and which could be sold freely.[4]

The growth of rural industry

More important was diversification into non-agricultural activities. In the collective period rural industry was officially confined to manufacture of inputs into agriculture. It greatly expanded after 1980. The real dynamic of the rural economy became the emergent 'township and village enterprises' (TVEs) that until the mid-1990s were to dominate not only the changing rural scene, but even the national economy. Between 1978 and 1987, rural non-agricultural employment increased by 50 million, about a quarter of this in manufacturing industry, and most of the rest in construction, transportation and commerce (Ho 1995). In a period of inflation at the end of the 1980s, credit to the TVEs was restricted and a substantial number failed. Counties seriously affected sometimes overcame the problem by successfully appealing to be reclassed as 'cities', thus gaining access to new sources of funding (Guo 2001). By 1995 rural non-agricultural enterprises as a whole employed 128 million people, and produced a third of the nation's GNP. Fairly heavily concentrated in more accessible regions of the country, and covering a very wide range of industrial and processing activities, they became the basis of rural income differentiation, and the means by which millions of rural workers escaped from poverty without (until the late 1980s) migrating on a large scale to the cities.

Most of the literature on the TVEs relates to the coastal provinces where there was extensive participation in industry for export from the mid-1980s onward. By 1991 exports by rural enterprises specializing in goods manufactured with a large input of labour reached 30 per cent of national exports (Ho 1995). Where there were TVE profits they have been of major importance in financing other aspects of rural development. A dissenting argument by Zhao and Wong (2002) held that investment in the TVEs has deprived agriculture of needed support and innovations, as well as taking large numbers of skilled workers from the fields. From 1995 onward there was rapid collapse of a high proportion of TVEs, and a drive to privatize those that remained viable (Ong 2006).[5] The initiative was taken mainly by the local party officials, often the same people as the managers of the TVEs (Li and Rozelle 2003). Although processing of local produce was a major activity, a proportion of TVEs neither bought from the agricultural sector nor sold to it and most of these have moved to the towns since the late 1990s. This transition has been accompanied by the growth of many small towns, and even some xiang, into urban centres with populations in the tens of thousands. Much the greater part of rural household income is derived from agriculture and the farmers' own sideline enterprises.

By the turn of the century, the boom in rural industry was over in all areas away from the cities and especially in the inland regions. In Sichuan, for example, many factories had been set up without a market for their produce being determined in advance. By 2000, in one county there were locally owned factories employing about 10,000 people, mostly producing a variety of products for the construction industry. Almost all were managed by former cadres. They had copied initiatives taken earlier in other areas, and were reported to be having major marketing problems in 2000 (Eyferth 2003). Like many others in the province they were poorly managed, with a great deal of waste, and ultimately failed (Ong 2006). Only an ancient high quality paper industry, revived after the Cultural Revolution, continued to do well, operating at family scale in small workshops scattered through the western Sichuan hills (Eyferth 2003).

Reconstructing rural China

Taking a long view of events in the 1955–78 period, some modern writers have been less inclined than earlier 'China-watchers' to condemn outright the excesses of the 'great leap forward' and the Cultural Revolution, or to treat the 1978 reforms as the liberating victory of pragmatism over ideological folly (Bramall 2006; Gray 2006; Kueh 2006). Labour-intensive investment in irrigation, drainage, infrastructure and rural industry were, after all, not so far removed from some Western ideas about rural development in the same period. Together with the mobilization of labour in the collectives, they did facilitate production increase and establish the preconditions for the rapid and diversified development that followed in the 1980s. The 1980s and 1990s also had their excesses and one of these was the extent of decentralization that followed the break-up of the collective system. Most public services became the responsibility of the lower levels of the administrative hierarchy (Ho 1995; Eyferth *et al.* 2003). Income from the TVEs funded these services, but only about one-third of villages in the country had collective enterprises, and something like a quarter was without collective financial resources. The consequences of this financial squeeze were passed on to the rural people, and particularly to the farmers who, together with their families, still accounted for 75 per cent of the national population in 1996, notwithstanding the advance of industrialization.

The 'peasant burden'

Thus arose what is described in China as the 'peasant burden' (*nongmin fudan*) on the farmers of predominantly agricultural communities. The regular taxes had to be supplemented by a range of additional fees and charges, augmented downward successively through county and township levels in order to meet the needs of enlarged local bureaucracies. These predatory additional charges became the major source for financing local governance (Yep 2004). Decentralization of service provision was inadequately compensated in central/local sharing of taxes, so that by the turn of the century the majority of China's 2,074 county governments, and also most township governments, were in debt (Yep 2004; Ong 2006). In total the burden of taxes and charges could amount to 13–20 per cent of farmers' often small income, worsened by the fact that delayed payment of taxes attracted interest and fines (Xiande 2003). Most of this, formerly charged to the collective, now came in the form of a demand to the individual holder of a household responsibility contract. Collection coincided with the sales to the Grain Bureau, and in a village closely studied by Xiande (2003) the farmer often got no balance for himself or herself from the first and second deliveries once the taxes had been deducted. In addition, while education and health services had been free under the collective system, there was now a charge for both. There were taxes to build schools, and then charges to use them. Commenting, Yep (2004: 69) opined that 'The Maoist strategy of modernization at the expense of agriculture survives in a disguised form in post-Mao years. The systemic discrimination against and disproportional extraction from the peasantry remains intact.' Some critics went further and commented that, together with the heavy alienation of peasant land for industry, roads built for private management and sale to enterprises with foreign participation, what had happened was something akin to the 'primitive accumulation' of Marx, by a new generation of 'gangster' capitalists (Walker 2006).[6]

Exactions began to trigger serious local protest in the 1990s, leading to a major policy shift. First, fees were consolidated into tax, while the inflated local government staffs were reduced. Then, with little warning, early abolition of the historical agricultural tax levied on agricultural land was announced and then took place ahead of schedule in 2006. It was by that time only a very small contributor to national revenue, but was of major importance to local governments. At the same time responsibility for health and education costs was shifted upward to the county and even the province (Li 2006). It was also reported that cash-strapped local governments deprived of taxation revenue were restoring fees in various guises. The size and diversity of China create constant problems in trying to administer national policy under radically different local conditions. Nonetheless, the excessive decentralization of the early reform period has clearly been reversed.

Reconcentrating the operation of land

Since the 1970s, there has also been some partial restoration of collective farming, both to meet collaborative needs among the farmers of more affluent areas, and as part of measures to ease poverty in the poorer areas. Although land continued to be held under individual contract, farming operations were sometimes jointly managed. By the mid-1990s, there was abundant evidence of a resurgence of group activities in some parts of the country (Kerkvliet and Selden 1998). At times de-facto re-collectivization has been spontaneous; at other times it was quite clearly guided. In the richer eastern regions it has sometimes become the practice for groups of farmers to rent out their holdings in large blocks to tenants who can use machinery and better organize irrigation. The tenants, not unlike the modern American share-croppers discussed in Chapter 4, meet the farmers' obligations while retaining enough to make the operation worthwhile. The farmers use their remaining land for specialized production, or take jobs in the nearby industrial areas. This is one of several ways in which the tiny holdings that resulted from the household responsibility distribution can be combined into parcels that are capable of larger-scale management. Another is for contracted land to be grouped into farms worked by only a few of the leaseholders using hired, often migrant, labour (Kerkvliet and Selden 1998: 52; Chan *et al.* 1984).

A growing number of farm households in all the more accessible and developed regions can now earn much more from off-farm work, principally in industry and commerce, than from the farms themselves (Zhou *et al.* 2002). Where the TVEs or their privately owned successors have survived, this income can be earned without migration away from home. Even in regions without nearby industrial opportunities, farmers have diversified into agroforestry production of tree crops, or even into forestry. This has happened despite the heavy tax and fee charges made in the southern provinces of China on the rural forestry sector by the state procurement companies (Yin *et al.* 2003) and the transit taxes that are still charged on agroforestry produce.

The continuation of 'command-and-control'

It is often argued that the post-1978 reforms have given Chinese farmers complete freedom of decision in how to use their land. Our Chinese colleagues in the PLEC project continued to insist that this was so in all their reports and papers. In qualifying their arguments, we recognize that we are looking at the modern situation from different points of view. Until the turn of the century, quota deliveries continued to be sought at the collective level to sustain food supplies to the urban areas under a long-standing national policy that sought to limit grain imports to no more than 5 per cent of national demand. Grain deliveries were required of individual farmers at the time of their tax payments, and the tax and quota systems were integrated, with cadres often working alongside Grain Bureau officials. In a region of western Yunnan where the quota was of sugar cane to supply poorly maintained factories in Nu Jiang (Salween) valley, the county authorities still required one *xiang* to plant 200 ha of its upland fields to sugar in the late 1990s (Dao *et al.* 2003). Deliveries are no longer sought from the higher land where the sugar content of cane is lower, but cane continues to be grown, in rotation with maize, on what would seem some very unsuitable land with very poor access. For these people, sugar remains the principal source of income.

An unknown number of rural workers, perhaps tens of millions, moved into the cities in the late 1980s, when the *hukou* (residency permit) regulations were partially relaxed. A further surge in the 1990s brought the total to around 130 million by the new century (Liu and Murphy 2006). With a large income advantage for the urban dwellers, the pressure of intended migration from the rural areas is never relaxed. Notwithstanding economic liberalization in the countryside, it has not yet changed what has become the major division within Chinese society, even greater than that between east and west. According to official statistics, urban incomes exceeded rural incomes by 2.5 times in 1978; in 2003, the margin had increased to more than three times (Li 2006).

Sustaining the supply of grain

Since the 1959–61 famine, security in grain production has been the central element in Chinese agrarian policy. The progress achieved since the 1970s is impressive. With only 7 per cent of the world's arable land, the system has fed 22 per cent of the world's population during a period of rapid increase. Despite substantial losses of farmland for urban, industrial and infrastructure development, and for re-forestation in more recent years, total production, yield per hectare and production per agricultural worker have increased in parallel for all crops (Tong *et al.* 2003). The contributory causes do indeed include the liberalization of farmers' decision making, much trumpeted by the media and most economists, but they also include Green Revolution crop-breeding improvements in wheat and especially rice. China's principal addition to the Asian Green Revolution was the development of exceptionally high yielding hybrid strains of rice, rapidly disseminated, at a cost, over a large part of the country and which are now most of China's rice crop. From the late 1990s, specialized grain farmers have received a direct subsidy, calculated on an area basis.

Even more important has been the massive application of chemical fertilizer which began during the collective period, but expanded greatly from the mid-1970s. Nitrogen is now applied in some parts of China with an intensity that equals any place in the world. As much as a third of Chinese nitrogen is made from the country's abundant coal resources, in the form of highly unstable ammonium carbonate, and there are large losses into the soil, water and atmosphere (Smil 2001). Applications of phosphorus and potassium remain well below levels common internationally. Unfortunately, not only has inefficient application led to large losses and severe pollution, but the effectiveness has rapidly diminished as the marginal yield of rice obtained for an additional kilo of fertilizer declined by half between 1979 and 1996 (Tong *et al.* 2003).

Modern society and the Chinese farmer

There are unresolved social problems. Under the post-collective system, farmers have to buy the fertilizer and pesticides individually, and they do this even in remote and underdeveloped parts of the country such as western Yunnan. One of the problems of the 'peasant burden' of taxation was that the first large instalment of the annual tax payment fell due when the winter crops were harvested, at the critical time when farmers needed to buy fertilizer for the summer crop. Since little cash remained once the taxes were paid, many farmers deferred payment, incurring interest and fines payable later in the year. This situation has been responsible for conflicts between farmers and the authorities (Xiande 2003). Such conflicts, and others over the compulsory and often uncompensated acquisition of land for the nation's rapid modernization, have continued, and by some reports at a rising level of anger (Walker 2006).

Farmers organizing on their own and innovating

Collaboration among Chinese farmers, like the farmers of many lands, has long depended on personal networks of cooperation that relied on kin, affinal relationships and friendly relations between neighbours. Because of the patrilocal pattern of post-marital residence normal in China, kin lived close, while affines normally provided links to other communities. Lineages and networks have been important in sharing work at peak seasons and providing assistance in times of hardship for centuries. They were very important in forming the mutual aid teams that followed land reform in the early 1950s, and in the first and voluntary cooperatives. After the ending of the collective farm system, most rural Chinese quite readily fell back onto their traditional networks. Sometimes these could be used constructively to foster the spread of innovations.

Wu and Pretty (2004) researched these networks in 50 *cun* in a county of Shaanxi Province, an upland area in the loess plateau of the northwest. They found considerable variation in the size of individual networks. Among the larger ones were what they termed 'innovation circles' where, usually under the leadership of one prominent individual, particular innovations such as greenhouses, grafting of apple trees, new cash crops or new agricultural tools were being introduced, tried out and spread. Villagers were often reluctant to adopt an innovation until they had seen it demonstrated, or until they had learned of its value from a trusted member of their personal networks. Rapid adoption, as of a new system of interplanting high value traditional rice with hybrid rice in Yunnan, is an exception to this generalization, perhaps because of the immediate financial gain made by the adopters (Zhu *et al.* 2003). Many of the *cun* were beyond the normal reach of the limited extension services, and some of the networks had developed stable cooperation around the new innovations. While they bore some likeness to the old mutual aid teams, they led toward a diversity of informal groupings, most effective where they could command resources (Box 7.1)

Box 7.1 The value of urban–rural connections

Wu and Pretty (2004) show how connections with migrants in the cities were sometimes important in providing financial support and, in one particular case, leadership. The engineer son of a Guomindang general who had been born in a remote village, was sent back there for re-education in peasant ways during the Cultural Revolution. Like a small minority among the many other urban elites who were banished to different parts of rural China, he returned to the rural environment after his rehabilitation at the end of the 1970s, to help his kin and their neighbours in overcoming problems that he could see would never be solved by waiting for government action. In this case, the last of the former woodlands had been cleared during the 'grain first' period, and even firewood was extremely short. He led a programme of cooperative re-forestation, using his own money initially, and then went on to other developments, including new cash crops and a road to link the village with the larger transport system. Even though he died in the village while this was still in progress, others took up his work. The village became a green island in a largely bare landscape, but also an island of relative prosperity in a poor region. Urban kin and affines of other villagers have provided support. The village leadership visited the agricultural authorities in the distant county town to obtain assistance, which they received with a willingness enhanced by surprise at being visited in this way.

Toward what future?

In modern China, effective local leadership has to operate both within the continuing command-and-control character of the party and state management, and to some extent also against it. This is not easy to do, even now when there is better appreciation of farmers' own ideas and methods than there used to be. Nonetheless, this is the way in which many changes have been brought about – working from below to obtain improvements. In the international media, it is very common to focus almost entirely on the top-down aspects of the system, but change can come from below as well as from above.

Yet continuing poverty remains a major problem. Large inequalities remain and even widen, despite the continuing egalitarian ethos; the gap between villagers and the decision-making centres remains enormous. About one in five among all the world's farmers is a Chinese farmer. The modern state is trying to adopt a more 'inclusive' approach to its huge rural population, but without slowing the frenetic pace of the nation's modernization, or abandoning the guiding role of the Communist Party. The scale of achievement is very large. By 2004, less than half the average farm household budget was spent on food, agriculture employed only 47 per cent of the national workforce and three-quarters of all rural households owned a colour television (Kueh 2006).

Notes

1 'Natural' villages were collected into 'administrative villages' long before communist times. The latter became the lowest level of the Chinese government structure. Hamlets within villages (*zu*), physically separate or not, remained distinctive as landholding groups, and in the collective period were often the basis of production teams.

2 Hinton's record of the land reform in Long Bow took more than a decade to be published. As a supporter of Mao's policies he suffered many restrictions after his return to the USA in the 1950s. When the Marxist-leaning Monthly Review Press finally accepted and published his manuscript it became something of a bestseller. But it was still some time more before he was able to recover his passport, and return to China (Anon 2004).

3 But he failed in the pursuit of his more educated and ambitious girlfriend, who finally tired of waiting for him to propose and instead seized an opportunity to achieve every progressive girl's dream – she went away to learn to become a builder of railway locomotives!

4 Sale of fruit, nuts and vegetables was freely possible in local markets, but if they were transported over any distance, the goods became subject to ad-hoc taxation levied at checkpoints along the highways. We saw this still in operation in 2002.

5 In collapsing, TVEs often put the local governments that had guaranteed their loans into serious debt. The management of the TVE, local government and the local credit-granting institution was often either the same people or their close working partners, while the lost money consisted largely of the savings of rural households. The problems of the TVEs and local governments created serious credibility problems for provincial and national governments. In closing down failed rural credit organizations, priority was given to the household creditors (Ong 2006).

6 Certainly there is an enormous amount of capital investment in rural China, not much of which has direct and obvious benefit for the local population. During the period since 1990, travel through China has resembled travel through an enormous construction site,

with only hints as to how it was all being financed. During the years in which PLEC was working in western Yunnan, Brookfield travelled several times between Kunming and the far west along what was initially the old 'Burma road', and later on a splendidly engineered and privately managed toll highway which also owned and operated all facilities along the highway. Large areas of productive farmland were alienated for this purpose, and it is unlikely that adequate compensation was paid.

8 The Periphery

From structuralism to neo-liberalism

Whereas the historical political economy of agriculture in the North Atlantic and the communist countries can be recounted in comparatively simple terms, the post-1930 history of the globally peripheral regions is more complex. Latin America and Australasia have been politically independent throughout the twentieth century; by a quirk of history one is classed as developing and the other as developed, though the economies of southern South America and Australasia have strong similarities. The countries of the tropical Asian and African regions gained independence only in the mid-twentieth century. The Green Revolution in the 1960s and 1970s happened most extensively in Asia; periods of military rule intervened mainly in Latin America and Africa; major warfare happened mostly in parts of Asia and Africa; and 'economic miracles' on a sustained basis have been characteristic mainly of Asia. Two forces have been general. These are the post-1945 drive for state-led economic development, converted during the 1980s into the neo-liberal project of globalism. This chapter can only be an essay on these events, though with specific focus on selected and contrasted countries.

As soon as metropolitan recovery from World War II was complete, one dominant theme in the developed world was to extend 'development' to the underdeveloped countries. Once it had been explained to a credulous world that the best indicator is growth of gross domestic product, or gross national product (GDP, GNP), major efforts were made to find out how to enlarge this statistical aggregation.[1] From the work of Keynes and his followers, it was known that growth demanded investment, which had to be funded by savings. By 1960 it was widely understood that to achieve a 'take-off' to sustained growth (Rostow 1960) not less than 10 per cent of GNP needed to be saved and invested in those leading economic sectors which would generate the highest and fastest return. In a literature summarized, in its own temporal context, by Brookfield (1975: 26–76) it emerged that manufacturing and services offered the best results, while agriculture and even infrastructure offered much poorer returns to investment.[2]

Structuralism

Another approach, significant to the argument of this book, arose in Latin America from the work of Hans Singer (1950) and Raúl Prebisch (1950). Singer had examined long-term trends in the 'barter terms of trade' between primary produce and manufactured produce, finding a progressive deterioration in the position of the primary producer. Prebisch, newly appointed as head of the Economic Commission for Latin America (ECLA), explained this deterioration in terms of the differing elasticity of demand for the two groups of produce, and (internationally) the structural contrast in the way in which the industrial and primary-producing economies were organized. These differences strengthened the advantages of what he called the 'centre', and we call the North Atlantic economies, during each prolonged economic downturn (Kay 2006).

Either way, the path for peripheral countries was clear. It was to promote industrialization for the domestic or regional market to reduce dependence on imported manufactures. This was, in fact, what several Latin American countries and others, including Australia and India, had been doing since the early twentieth century. Primary produce exports would finance this industrialization, and they could be taxed to encourage investment in industry. Shortage of capital led to efforts to attract foreign direct investment, expanding the 'branch-plant economy' that was already well developed in such North Atlantic 'semi-peripheral' countries as Ireland and Canada. For populous developing countries, it seemed that labour could be drawn out of agriculture into industry at a cost only a little above the agricultural wage. Agriculture itself would not suffer since it was already burdened by an excess of potential labour (Lewis 1954). In effect, though Lewis did not say so, the farming population of developing countries was envisaged as the location of Marx's 'reserve army of the unemployed'.

The ECLA group also saw it as essential that rural purchasing power be enhanced, notwithstanding persistent weakness in the price of primary produce exports. As servants of Latin American governments, they were initially reluctant to propose any large-scale redistribution of land as a palliative, but in 1961, in the aftermath of the Cuban revolution, this became briefly popular not only with national authorities, but even with the increasingly involved USA. The surveys coordinated by the Inter-American Committee for Agricultural Development in the short period when land reform was widely and internationally favoured reveal the dimensions of inequality.[3] This is summarized in Table 8.1 for seven countries for which reasonably comparable, if somewhat subjective, data were collected. The dominance of large estates (haciendas) is very evident. Called 'multi-family' and 'large multi-family' farms by the survey, most were worked by labourers or tenants. The leading landholders controlled several such estates, while in each country from 23 to 60 per cent of all farm workers owned or rented no land (Barraclough and Domike 1970).

All this thinking took place in the new context of national planning, which had become popular in almost all countries during the World War II. In no small measure, this arose from admiration of the success of the USSR in building its industrial economy during the 1930s to the strength which enabled the country to

Table 8.1 Farms and land in seven Latin American countries in the 1960s

Type of farm or holding	Farms(%)	Land area(%)
Sub-family farms	51.7	2.5
Family farms	30.1	21.0
Small estates	15.6	31.0
Large estates	2.6	45.5

Note: Aggregated from country data in the final report of the Comité Interamericano de Desarollo Agrícola (CIDA 1971). The countries are Argentina, Brazil, Chile, Colombia, Ecuador, Guatemala and Peru. The classifications seek to avoid problems arising from using simple farm size in the context of large ecological differences. Sub-family farms are insufficient to meet the minimum needs of a family; family farms have land sufficient to support a family and employ their labour through the year; small estates employ non-family workers, but are not large enough to require a managerial staff in addition to the owning family (i.e. small haciendas); large estates, or *latifundia*, are large enough to support complex division of labour and an administrative staff (i.e. large haciendas). Based on a table in Brookfield (1975: 152).

resist and defeat the German invasion. Five-year development plans, inspired by the Russian example, were initiated in most of the new Asian and African countries created by de-colonization. Industrialization called for use of most of the capital that the developing countries could muster, and since there was not enough, necessary capital was also provided by the substantial aid transfers that began to flow in the 1950s. It was national planning, more than industrialization, which was opposed from the North Atlantic region, where it was seen as possibly leading to exclusion of transnational enterprises (Coatsworth 2005). We now briefly sketch events in three major regions during the development drive of the period up to the 1980s.

Latin America

The distinctive landholding and labour history of Latin America marks it out from other developing regions, and this history is a necessary part of the story. Under the Spanish in the seventeenth to nineteenth centuries, large estates (haciendas) had crystallized out of the allocations of land to powerful men. Even in the densely populated regions there were a fair number of haciendas larger than 10,000 ha. Already in the seventeenth century the labour force of such estates was obtained and retained by 'debt peonage'. In Spanish America all but the slaves were notionally equal in law, but 'masters paid all their laborers' tributes, advanced them money, clothed them, gave them medical attention when they were ill, and thus kept permanent debts accumulating' (Chevalier 1952: 285). Not only were the workers bound by these debts, but their sons inherited the bondage. Later, it became common for trading stores to be set up in large haciendas, where goods were available on credit, and the debts were further increased. Bondage was even more total where chattel slavery prevailed. In Brazil slavery did not end until 1885, and in the rubber-collecting regions of the Amazon basin a form of debt-bondage barely distinguishable from slavery continued into the twentieth century.

Although all but a few areas around the Caribbean Sea had achieved independence in the first third of the nineteenth century, the skewed pattern of landholding did not change, and nor did the skewed relations of power that went with it. There were a number of uprisings, but none enjoyed significant success before the Mexican revolution of 1910–17. In the later nineteenth and early twentieth centuries, commercial agriculture grew substantially, both for export and to meet growing urban demand, and in the process many haciendas enlarged their holdings of land and labour by absorption of indigenous land. In Brazil, where no significant indigenous peasantry survived, large-scale government-funded immigration from Europe supplied a new population of workers for the expanding coffee industry, on bonded terms that gave them some land, some paid work and some obligatory labour (Martins 2002).

Until the 1950s, most of the substantial amount of rural warfare in Latin America involved small farmers (campesinos) as soldiers for factions among the landed gentry; one part of the Mexican war is the only outstanding exception (Chapter 11).[4] Communism eventually became a force in the prolonged Colombian warfare, although most of this was the interminable conflict between long-established political parties, as readers of Gabriel Garcia Marquez's famous *One Hundred Years of Solitude* will remember. Only by the retroactive decision of its leaders did communism become the winner of the Cuban conflict, in 1959. The Bolivian revolution of 1952 was an uprising of rural and urban workers, and although it destroyed the power of the great haciendas, it led only to a very modest redistribution of land in the crowded Andean region (Garciá 1970). Perhaps the principal beneficiaries were those who moved onto the Santa Cruz plains east of the mountains, and the Mennonite colonists and later Brazilian migrants who soon outnumbered them; they acquired holdings of adequate commercial size (Hecht 2005). Ché Guevara's attempt to inspire more radical revolutionary change in Bolivia in the 1960s was unsuccessful for the Bolivians and fatal for himself.

More lasting events took place in Peru, where attempts to reoccupy land seized from the indigenous Quechua farmers of valleys in the Cuzco region were supported by a Peruvian leader with Trotskyist world-revolution ideas, Hugo Blanco, who was captured but survived to be later released (Blanco 1972). The Quechua farmers did get back some of their land, and there was a more enduring overturning of estate-owner control under a military regime a few years later. Also significant for the future, though more immediately suppressed after the military takeover in 1964, were the Brazilian '*ligas camponesas*' (peasant leagues) formed among coffee and sugar-cane workers being displaced from their tied holdings to become proletarian labour in Brazil.

The limited results of national efforts soon led to the gloom of the 'development of underdevelopment' period of thinking. Underdevelopment was seen to be a direct consequence of the deep and lasting penetration of capitalism, using, adapting and reconstructing the colonial and even pre-colonial institutions of Latin American society and economy. This view, which from the beginning underlay a good deal of the efforts to find an import-substitution escape from dependent underdevelopment, was particularly well expressed by two authors, one writing at the beginning of modern times (Mariategui 1928), and the other in the years when much of Latin America was already falling under the sway of right-wing military governments (Frank 1967). It was a situation that left Latin American campesinos in a subordinate relation to the landowning class, and in turn to the growing business and political groups (the 'national bourgeoisie') who were increasingly the clients of international capitalism. By the 1970s, the greater part of Latin America had fallen under the rule of a set of remarkably repressive military regimes, and it is this period, and its sequel, that marks the real hinge to the present in the continent as a whole.

Southeast Asia

By the end of the 1960s, the pride of place hitherto given to Latin America in world thinking on development had already begun to yield to Asia. At first, the new attention to Asia was very specific to agriculture, for in the 1960s the new high yielding rice and wheat were first used on an extensive scale in Southeast Asia and India. We discussed these events in Java and northern India in some detail in an earlier book (Brookfield 2001: 218–37). In Indonesia, and in Java especially, rice was central to national politics from 1966 to 1986; in few countries have farmers ever been so comprehensively instructed on what they should do. In Malaysia, there were similar attempts to build national self-sufficiency, coupled with import-substitution industrialization. Both Indonesia and Malaysia embarked on large-scale land development schemes, though with different emphasis. The Indonesian trans-migration schemes had the object of resettling people from 'overcrowded' Java in the outer islands, and set out to create a largely subsistence-based peasant population. The Malaysian variant aimed to create commercial farms producing rubber and palm oil for export. The Malaysian drive overreached itself. The Federal Land Development Authority (FELDA) was by far the principal among the parastatal bodies involved, and the main agency for the rapid development of oil palm in place of rubber. By 1990, FELDA hoped to have cleared 8,000 sq. km, and to have settled a million people, but this was not to be. Between the early 1960s and the late 1980s the combined effect of a near-doubling of the agricultural land area and a static population of legal rural workers in Peninsular Malaysia had led to a massive reduction in the effective density of rural population (Brookfield 1994). Malay recruits for the schemes were drying up, and the land could not be worked without the services of more than a million illegal immigrants, overwhelmingly from Indonesia.

A short recession in the mid-1980s exposed the unsustainability of the situation. Some 40 per cent of FELDA land was without clients, and in 1991 it was announced in the Sixth Malaysia Plan that these schemes would be operated on a plantation basis, meaning that they would use the immigrant workers to earn income for the large agribusiness corporation that FELDA had become (Malaysia 1991). And yet there remained poverty in rural Malaysia, and most of it was experienced by village-dwelling ethnic Malays growing rice and rubber (Corner 1983). When a poverty-line income was first determined, it was found that 88 per cent of rice farmers, and 65 per cent of rubber smallholders and mixed farmers were living in poverty in 1970 (Malaysia 1976).

Together with Singapore, with which it was constitutionally linked until 1965, Malaysia led the way in joining Southeast Asian economies to the east-Asian group of 'newly industrializing countries'. All this happened since around 1960, when a number of commentators viewed the prospects with considerable concern, not to say gloom (e.g. Silcock and Fisk 1963). Import-substitution industrialization had made considerable progress, but was beginning to shift into the export-based mode that was to enjoy such striking success. In a well-documented but little regarded book, Yoshihara (1988) found development in the region dangerously dependent on government sponsorship with international finance, seriously beset with rent-taking and speculation, and with its local capital very heavily invested in non-productive areas such as the spectacular high-rise buildings in city centres. Especially in the state sectors, he found much inefficiency, and generally an almost unrelieved dependence on foreign, especially Japanese, technology. These problems, and especially the high exposure to foreign-currency debt, were to come home to roost when the regional financial markets collapsed in 1997. All this notwithstanding, in the case of Malaysia the policy of accelerated export-based industrialization has succeeded to the extent that manufactures now supply four-fifths of merchandise exports. Some of these manufactures use Malaysian timber, palm oil and rubber, but a significant proportion has very limited backward linkage within the national economy, and depends primarily on an adaptable but relatively inexpensive workforce which, as in other regional countries, is young and still largely of rural origin.

Sub-Saharan Africa

In Africa, agriculture in the colonial period had two assigned functions: to produce export crops that, like mineral exports, would earn foreign exchange; and to nurture the labour that would work on the land of colonists, in the mines and towns, and support their families who remained behind. This labour migration was substantial, especially in South Africa where all that remained to the Africans were reserves that in 1945 occupied just 9.6 per cent of the area of the country. A government commission in the 1950s found that in high and low rainfall reserves practising mixed farming, 43 and 67 per cent respectively of African family incomes were supplied by wages earned in the European-controlled areas outside the reserves (South Africa 1955).

In most of the rest of sub-Saharan Africa, cash-crop income added to subsistence production was more important than labour migration. In some countries the adoption of cash crops had been rather fiercely enforced by the colonial authorities, but by no means everywhere. There were important local initiatives, of which the earliest and most famous was the establishment of the cocoa industry in southern Ghana by farmers settling a largely vacant forest area from an adjacent upland ridge that was their homeland (Hill 1963). After World War II, the drive to expand production of staple export crops was intensified, and the same policies were continued after independence. Support to agricultural production in Africa never achieved the levels it attained in several Asian countries in the 1960s and 1970s, but there were substantial results, especially with cotton in the francophone countries of West Africa (Bassett 2001).

Despite subsidies, rewards remained meagre except to the largest farmers. An International Labour Organization (ILO) mission to Kenya at the beginning of the 1970s, when 90 per cent of that country's population still lived in rural areas, found only a small minority of farmers to be in the top three income brackets of the nation, while the lowest three income brackets – mostly deemed to be in poverty – consisted principally of rural smallholders, farm workers, the unemployed and landless, and the informal sector (ILO 1972). This was after the break-up of almost half the large farms that before independence had occupied four-fifths of the country's best land for the benefit of some 4,000 European settler families. Most of the benefit of this redistribution went to only about a fifth of the nation's smallholder farmers. This sort of income distribution was not uncommon.

The ILO (1972) mission to Kenya is justly famous for its discovery of the urban 'informal sector', though less famous for its largely ignored subsidiary discovery that a large sector operating outside formal sanction or regulation is also present in agriculture. The 'informal sector' is not all that modern, and by no means a characteristic only of Africa. It became significant in Africa as both industry and agriculture failed to satisfy the demand for rewarding incomes during the 1950s and 1960s, before the economic crises that followed in the 1970s and 1980s.

This was not due to any shortage of development specialists from the developed countries, who sought to set up industries and transform agriculture. The way in which most of them operated, however, was paternalistic at best. An example that can stand for many was an effort to modernize agriculture in Senegal through irrigation and the promotion of commercial rice production, by individual farmers working on settlement blocks. Adrian Adams (1986) tells how, during the 1970s, attempts by farmers to recruit assistance for their own proposed enterprises were captured by international experts working through state agencies. They imposed external designs that included the breaking up of village communities and their well-adjusted systems of mixed subsistence and commercial farming. While some communities followed directions, others obstinately retained the local organizations they had set up, and continued to make their own decisions. Unsurprisingly, they were starved of technical and financial aid. The development drive sought to jump in one leap from an imagined pre-colonial, pre-capitalist society to a modern society of individuals administered for their own good, missing on the way any recognition of the uneasy, adapting present.

One of the reasons for poor returns to export-crop farmers was the establishment of marketing boards or their equivalent in the late-colonial period. Marketing boards in the developed countries, mostly set up during the crisis years of the 1930s, legally belonged to the producers, whose representatives controlled them. Those set up in the developing countries have all been controlled by government.[5] They were set up to manage trade in export crops grown for processing and consumption in the developed metropolitan countries, and for this reason they sought to keep prices down rather than to raise them. Independent governments did not reverse this trend, because the boards were seen as a means of taxing agriculture in order to generate funds for expansion of manufacturing, and to support government itself. The effective 'tax', comparing national buying prices with the free-on-board prices prevailing in the world market, was often of the order of 50 per cent or even more. The boards not only monopolized purchase, but also monopolized storage and export.

After independence, many developing country governments set up additional marketing authorities to control the purchase and internal distribution of basic foodstuffs, in particular the grains. Again, the purpose was to hold down prices, this time for the benefit of the employers of urban workers by suppressing upward pressure on wages. Although the new boards had the sole legal authority to purchase commodities and engage in trade, they often lacked the means and skilled personnel to do this effectively. The cost of maintaining, or contracting, purchase and distribution networks and storage facilities proved very high. The inadequacy of the new authorities also meant that in a majority of countries parallel private markets, or black markets, continued to flourish.

By the 1970s, state-managed development had not only failed to produce the anticipated results; it had also led to deep indebtedness. To this problem were added continuing anti-colonialist warfare in southern Africa and an El Niño-precipitated famine in the Sahelian region, the first of a number that would strike African farmers in the next 30 years. It has become the habit to regard sub-Saharan Africa as the 'basket case' of world development. Civil war, never absent from Africa since the 1960s, still continues in several countries in 2007. All this left rural Africa with little protection against a set of political forces that were brewing in other parts of the world (Chapter 6).

Neo-liberalism: resurgence of a discarded doctrine

By the 1970s the import-substitution industrialization drive had reached its limits, even in countries such as Brazil where it had been most successful. The model of state-led economic management was running into difficulties in all parts of the world in which it was applied, in developed and developing countries alike. For a few years, there was call for a 'new international economic order' which would undo discrimination against the primary-producing developing countries. These changed conditions gave new scope to what had been a small minority of 'neo-liberal' economic theorists who saw state participation in production and trade as pernicious, and believed with all the fervour of their nineteenth-century free-trade predecessors that the free market is the proper way in which to manage human affairs, both within and between nations.

There were important corollaries. Operation of the free market required unfettered private control over factors of production and trade. Private property, in the undivided and absolute sense used by Marx and nineteenth-century liberals, was central. The more modern notion of property as a divisible bundle of rights (Chapter 3) was not so much opposed as ignored by the neo-liberal economic philosophers. Comprehensive individual rights should rule out a role for anything more than a minimal state. Yet governments were called on to remove all 'distortions' from operation of the free market's 'hidden hand' (Gasper 1986). They were also required to manage their currencies in ways that would support trade, and attract foreign investment. The conditions to enforce these policy shifts were not readily obtainable in countries where the state was securely embedded in the whole operation of society and economy, but international financiers could impose them on weak developing country governments when they were in need of assistance in times of economic crisis.

Neo-liberalism was not just nineteenth-century liberalism reborn, although often presented as such by its advocates. The liberal creed which arose in the 1830s had rested on three pillars: a free market in the factors of production, especially labour and land; free international (and internal) trade; and a sound currency, then based on the gold standard (Polanyi 1944). Neo-liberalism had to confront controlled markets, highly contestable international trade and unstable currency following the collapse in the early 1970s of the system of international currency management that had been created in the aftermath of World War II. As in the 1830s, the laissez-faire they sought to achieve had to be created and planned. It was not easy. The story is complex, and defies simplification even at regional level, as in Latin America or sub-Saharan Africa – although good efforts have been made to do so. Our approach is therefore selective, focusing principally on the impact in two countries that received early 'shock therapy', Chile and New Zealand.

Chile

Events in Chile in the 1970s provided the neo-liberals with an early opportunity to demonstrate the virtues of their approach to a world disenchanted with state-led and regulatory approaches. Chile was late in embarking on a socialist development path. After 1964, and then more radically after 1970, a land reform which expro-priated the large estates was joined to policies involving income redistribution and nationalization of the mines, banks and other key economic areas. By the end of 1973 almost all haciendas had been expropriated. Most of them operated on a cooperative basis under government management which, to many of the campesinos, resembled their former relationship to the estate owner (hacendado). Production was broadly sustained, but significant improvements to urban and industrial wages led to larger demand which had to be met by imports. Serious shortages provoked unrest among the middle class.

A mixed agricultural economy was planned, with a significant shift to higher-value exports rather than production for the national market – the 'import-substitution' pattern then proposed for agriculture in other countries. There was also to be major reconstruction of marketing, processing and input provision (Conchol 1973). Such reconstruction had not begun when, in late 1973, the Popular Unity government was overthrown by a violent military coup d'état. The new government returned only 33 per cent of the expropriated land to the original owners. They partially implemented the reform by allocating 41 per cent of the land, mostly cooperatively managed under the two previous governments, to individual campesinos, as *parceleros*, in units large enough for economic farming. Some 16 per cent was auctioned to companies and individuals, and the remaining 9 per cent was allocated, often without payment, to the armed forces and public institutions (Bellisario 2007). The cooperative experiment was ended, and those workers who did not receive land were dumped onto the employment market in a period of severe depression.

The military government was advised by American neo-liberal economists of the school bred in the University of Chicago (hence '*los Chicago boys*'). From 1974 until 1990 the Chilean economy was operated on a basis as close to neo-liberal principles as was possible under authoritarian control. Import substitution gave way to export orientation, both of minerals and timber from the land, and of a new suite of crops from the farms – very much as the socialist regime had proposed. With substantial external financial support, significant economic growth was achieved in the 1980s. The long-term consequences for the farmers were very mixed. Some *parceleros* had engaged in the rising export industry of commercial fruit farming, but not all had received enough land for full commercial production. They fairly quickly fell into a debt relationship to the buyers of their produce, to whom they became contracted on increasingly rigorous terms. Many lost their land and became proletarian workers (Murray 2002; Gwynne 2003). The net result of the Chilean experience, under three very different governments, was conversion of a classic hacienda-based agrarian system into a rural economy based on capitalist principles (Bellisario 2007).

'*Structural adjustment*'

Financial problems grew much worse when internationally mobile capital was greatly augmented by the flood of petro-dollars in the later 1970s, undermining the ability of nation-states to manage financial flows across their boundaries (Chapter 6). The real 'new international economic order' was now seen to be managed not between states but across them by transnational corporations and financiers. With inflation and rising interest rates, indebtedness in many countries rose to levels at which it could not be serviced. The collective indebtedness of Latin American countries threatened to wreck the whole global financial system if its largest countries were to have defaulted simultaneously.

International finance now had its opportunity to impose neo-liberal conditionality on all but the strongest of these countries. Contrary to calls a few years earlier for a more equitable new economic order, they were required to lower or eliminate barriers to imports while not receiving corresponding concessions from North Atlantic trading partners. Under pressure from the conditions attached to all new international finance, the weaker developing countries had to drop their own subsidies, such as they were, and to dismantle or downsize government involvement in production and marketing. They were obliged to devalue their currencies or allow them to float, reduce government employment, privatize state enterprises, adopt stringent conditions of fiscal discipline, and impose charges for public services such as education, health and agricultural extension. Above all, it was seen as imperative to 'get the prices right' to attract and retain private-sector international foreign direct investment. In sum, they were required to drop most remaining vestiges of the largely socialist-inspired development policies with which many of them set out in the 1950s and 1960s, and instead to join the 'race to the bottom' in pricing their economies to attract foreign investment. The enforced new policies, which were developed in financial institutions headquartered in Washington – hence the term the 'Washington Consensus' – were subsequently adopted by most of the multi-lateral and bilateral aid funders in other parts of the world (Rodrick 2006).

In the North Atlantic countries themselves, farmers retained the state support that had become established since the 1940s and, as we saw in Chapter 6, their political backers were even able to negotiate improvements in this support as late as the early twenty-first century. Although the conditions attached to this support began to change, in 1985 in the USA and in 1992 in Europe, the support itself nonetheless remained in place through the first wave of neo-liberalism. Away from these countries there was no comparable protection from the new forces, and with greater or lesser reluctance – sometimes with no reluctance at all – many governments imposed the new conditions on their people, including their farmers. Along with their international financial backers, they hoped to see the market deliver the freely moving development that state-led policies had failed to produce. Authoritarian and democratic regimes alike have followed this new 'shining path' – a broad international consensus which only shared with the Maoist so-named 'shining path' (*sendero luminoso*) rebellion in Peru a determination to see the size and role of government reduced!

The neo-liberal transformation of Chile continued with few basic policy changes, except in the social sphere, after the dictatorial Pinochet regime was replaced by elected successor governments. In the much more dynamic situation in Brazil, the military government had strengthened the position of the landowners and savagely oppressed the socialist movements of the early 1960s. Yet they also enacted a major social security fund for rural workers and small farmers, administered through Brazil's corporatized union movement which the regime was seeking to co-opt (Welch 2004; Harnecker 2002; Martins 2002).[6] Here, it was left to subsequent democratic governments to carry out most of the neo-liberal agenda, in an uneasy relationship with modern social movements among both farmers and workers. The pressure to liberalize trade, while at the same time protecting the economy against subsidized imports through countervailing duties, came largely from the large farmers' and landowners' organizations, which acquired greater influence after the end of military rule (Helfand 1999). Although the great debt crisis of the early 1980s exposed Brazil to external pressures, transmission of change was mediated through a complex system of internal political forces. Generally, and in most developed as well as developing countries, 'market fundamentalism' was applied in a more nuanced manner than in Chile. It thus curiously happened that the next 'shock' imposition of the neo-liberal agenda took place under a social democratic labour government in wholly democratic New Zealand.

New Zealand (and Australia)

While New Zealand is classed as a 'developed' country, it had and still has a developing country economy heavily dependent on primary produce exports, and principally on agricultural exports. As part of a larger pattern of 'reform' to escape their problems of stagnation and indebtedness, the New Zealand government set out in 1984 to expose their farmers to the world market in very short order. They had already lost the privileged market in Britain for meat, butter and some other produce when the British adhered to what later became the European Union in 1973. Temporary relief was then given to farmers through 'supplementary minimum prices', especially important to sheep farmers, but in the space of less than a year from 1984 they lost not only this easement but also more than 30 other subsidies. Advisory services were privatized. All this happened during a period of newly floated currency and rising interest rates (Cloke 1996; Smith 1994; Smith and Montgomery 2003). Also affected, and in Australia as well, was credit to farmers. It had been available at concessional rates for most of the period since 1950, during which banking and credit were highly regulated in both countries. Credit had been available to finance a period of major expansion in agriculture and livestock production. These concessional loans were no longer available. Rural banks were privatized, but in order for this to occur, government had to assist the writing off of a large body of non-recoverable debts. In Australia, some state governments, closer to their rural constituents than the national government, picked up rural credit, but on commercial terms. Farmers' action groups had some effect in negotiating better terms from the banks during a severe debt situation in the early 1990s (Argent 2002).

It was anticipated that there would be substantial restructuring in New Zealand, and that 'less-efficient' farmers would be forced out of the industry. There were losses, but the majority of the farmers who failed financially were among the highly indebted, and included many younger farmers whose production methods were thoroughly efficient (Smith 1994; Smith and Montgomery 2003). The entrepreneurial farmers who filled their place included many newcomers who worked their farms hard to make a profit with little concern for environmental deterioration, intending to sell after a few years (Johnsen 2004).

Much less happened in Australia. The main changes were in marketing, from which national and state governments progressively withdrew. In the dairy industry marketing was not deregulated until 2000 and farmers' organizations had been able to negotiate some improvements in milk prices to take account of the increasing cost-price squeeze. By 2000 most milk processing and marketing was in the hands of a very small number of companies; collective bargaining with the processing companies was no longer possible, being construed as tampering with market forces (Anderson 2004). Yet Anderson (2004) shows how, as in New Zealand, farmers who were already wholly commercial could still retain traditional ideals and practices, and survive.

Subsequent developments in New Zealand agriculture have, on the whole, been positive, with a major increase in dairy exports to new markets displacing wool in importance. New Zealand's numerous dairy cooperatives have now concentrated into a single private company. A formerly small and undistinguished wine industry has grown very substantially as an exporter of quality wines. Growth was well behind that of Chile in the same post-'reform' period, and control has fallen mainly into the hands of two transnational companies. By contrast in Chile enterprising local wine companies have built up a major international business (Gwynne 2006). On the other hand, in New Zealand there has been no strong tendency toward concentration of land, or proletarianization of small farmers, as there has been in Chile. Many New Zealand farmers or their wives have taken off-farm jobs, and there has been on-farm diversification. The departures from the land mainly took place in the early period, and there has been less subsequent attrition. All this has taken place with the minimum of aid from government, and with a sharp decline in funding for agricultural research. Even in a country with every possible comparative advantage for agricultural production, successive governments since the 1980s have followed most international economic theorists in treating agriculture as a 'sunset industry' – even using this term at times.

The New Zealand dependence on its farm exports and the resilience of its farm families underlie the positive response that a great many farmers have made. W. Smith (1994) and especially Johnsen (2003, 2004) have drawn attention to aspects of life under sudden neo-liberalism that most writers have ignored. Stress imposed by the hard times of the mid-1980s had lasting physical consequences for many farmers, still enduring 20 years later. Farmers had to dispense with hired labour, undertake their own fencing and even shearing, grub weeds by hand, reduce livestock numbers and heavily cut down on chemical inputs. The length of the working week was substantially increased. The share of farm work and management undertaken by women increased greatly. Surprisingly, the reduction in external inputs did not seriously impact on production, but lack of investment into land maintenance did lead to an increase in erosion, although the relative shift from sheep to beef probably improved grassland management. The families, who kept their farms intact and succeeded in adapting their activities to the new situation, even returning to profitable operation, are naturally very proud of their achievement. But all this did not lead to higher farm incomes. It rested on heavy Chayanovian self-exploitation.

Stark neo-liberalism in retrospect

Viewing the period since 1980 in perspective, many observers of New Zealand and Chile have taken the measure of success as justification of neo-liberal, free-market policies. Others have pointed to longer continuities. In New Zealand, Smith and Montgomery (2003) have pointed out that all the trends that have succeeded since 1984 were already under way in the 1970s, and that the normally delayed response to earlier research and development has been at play in the productivity improvements; its benefits are now naturally slowing down as a consequence of the 1980s cuts in agricultural research (Johnson *et al.* 2006). In Chile, we noted above that the idea of a shift to export-oriented farm specialization was already present in the early 1970s, and even 1960s, years before it really took off in the 1980s. Some of the most specific technical improvements, in dairying in New Zealand, in viticulture and wine-making in both countries, and in fruit farming, are the product of a much wider international technical revolution that also has earlier roots.[7]

Neo-liberalism hit most developing countries of Africa and Latin America in a period when the received prices for their agricultural exports were in severe decline, in part because of the success of Asian countries in enlarging production of the same commodities. At the same time, the import substitution manufacturing industries became increasingly unprofitable, due largely to the high cost of inter-mediate goods. Although urban wages remained higher than rural incomes, they depended increasingly on skilled employment, not easy for unlettered rural migrants to get. Between the 1960s and 1990, migration to the towns in Brazil was increasingly to the shanty towns where informal sector activities were the only available employ-ment. All this left dispossessed sharecroppers, and farm workers displaced by mechanization, with few options. It led to the strong counter-movement of the landless rural workers' movement (*Movimento dos Trabalhadores Rurais Sem Terra*, MST), which we discuss in Chapter 11.

One of the strongest demands of the neo-liberals has been withdrawal of the state not only from production, but also from marketing. Under 'structural adjustment' in the 1980s many of the national marketing boards have been either wound up or reduced to the role of providing buffer storages, and of acting as buyer of last resort. In the case of Indonesian rice, these functions continue to give the marketing board the role of price leader, and many of the private dealers remain prone to government interference in their activities (Barrett and Mutambatsere 2006). In Madagascar, where marketing conditions in the 1960s were discussed at some length in Chapter 5, socialist arrangements had not worked well in the 1970s. In 1983 the indebted regime followed its bankers' advice (instructions) and liberalized the system, anticipating entry, or re-entry, of private enterprise. The first priority was to improve marketing of rice. This was not easy. Powerful local authorities delayed change in rice marketing by limiting the number of licences given to collectors and by setting up road blocks to shut out private transporters. By the later 1980s improvements in producer prices had prompted expansion of production in more productive areas (Berg 1989). Closer to Antananarivo, where a majority of rice-producing families needed to buy rice in the pre-harvest season, wildly fluctuating prices prompted some violence directed mainly at the larger wholesalers and long-distance transporters, who were often ethnically Asian. Not until the 1990s was the new system securely in place.[8]

Writing more generally about deregulation of internal marketing, Barrett and Musambatsere (2006) argue that rapid privatization exposed the weaknesses of the indigenous trading, processing, storage and transport systems. For Madagascar, comparison with Donque's report summarized in Chapter 5 might show there were good reasons for public regulation, but might also suggest that, for all its inequities, the old marketing system worked and was progressively adapting to changing conditions.

Winners and losers

Many national economies were enhanced by the restructuring brought about in the 1980s and 1990s, so that there were many beneficiaries. Many New Zealand family farmers, and a certainly smaller proportion of Chilean *parceleros*, adapted successfully to being thrown onto the open market by adopting new products and practices, and by diversifying their household economies so as to earn money off-farm. But increasing farm household diversification was not initiated only by the shock of neo-liberalism. In most developed countries, rising production for essentially static markets, coupled with the rising cost of inputs, has led to price weakening; it cannot be attributed in any meaningful way to neo-liberalism. If we are to look for the true losers from neo-liberal policies, we need to look among the poorest in the developing world. A specific case in southern India will illustrate better what has taken place than any general discussion.

Southern India

Telangana is an inland region, of more than 25 million people, within the southern Indian state of Andhra Pradesh. Since the 1980s, when Indian economic recon-struction began, its agriculture has been transformed by increased participation in the commercial economy, producing rice for the market and cotton for the national textile industry. Expansion of agriculture has been facilitated by the belated adoption of high yielding varieties and technology and an expansion of well-irrigation. In common with the rest of India, it has been subjected to neo-liberal policies since the early 1990s. Liberalizing markets and reducing subsidies were changes aimed at improving the efficiency of agriculture and leading to better growth. This indeed seems to have happened. Regional agricultural production, dominated by rice and cotton, increased at a rate of 4.7 per cent per annum in spite of a general agricultural depression, and in the context of declining trade prices for both rice and cotton. But to see more closely what happened, we draw on a paper by Vakulabharanam (2005), based on a study of four villages in the regional context.

During the 1990s even the large farmers (in Telangana those with more than 4 ha of land) experienced a slight decline in their welfare, as measured by consumption. Small farmers and the landless suffered worse. Between 1993 and 2000, the number of marginal farmers (with less than one arable hectare) fell sharply, while the number of agricultural labourers increased by almost 9 per cent. Increase in market involvement began under quite favourable conditions in the 1980s, leading to progressive decline in the cultivation of the nutritious 'coarse grains' used for subsistence. When markets were opened in the 1990s, received prices for the cash crops ceased to grow, while the cost of inputs – fertilizers, pesticides and the electricity needed for well-irrigation – increased substantially. The cash crops, especially cotton and chillies, demanded more labour. A compar-ison of input costs against received prices for cotton demonstrated that in most years since the mid-1990s farmers have barely been able to keep ahead of rising costs.

Under these trying circumstances, farmers increased output by working longer hours for more days in the year. In the study villages, the average working day increased from 8 to 12 hours between 1985 and 2000, and the working year from 240 to 300 days. This Chayanovian response unfortunately has not shielded the peasants from increasing dependence on the merchants, who provide credit and require the crop to be sold through them. Government sponsored micro-credit through women's cooperatives has the burdensome aspect that failure to pay on time reflects severely on a woman's standing, while being grossly inadequate for production needs. Increasingly therefore, farmers are turning to the moneylenders, who charge from 24 to 60 per cent interest, depending on the urgency of need, against the 10–15 per cent charged by the formal credit sector. Merchants charge similarly high interest rates. Farmer suicides, uncommon before the 1990s, totalled more than a thousand between 1998 and 2002 alone. As Vakulabharanam (2005) explains, spectacular modern agricultural growth in Telangana is driven by distress, and generates distress. Only the merchants and moneylenders do well.

Neo-liberalism with a kinder face: poverty reduction

The bitter medicine of structural adjustment yielded only limited results. Private-sector foreign direct investment obstinately refused to spread beyond a minority of developing countries; political reactions mounted. When the Southeast Asian economic bubble collapsed so spectacularly in 1997, there was a resurgence of pro-poor thinking in the Washington-centred international development management industry (Craig and Porter 2006; Rigg 2003). The Southeast Asian collapse, short-lived as it was, was psychologically critical. The rapid post-1970 economic growth and substantial poverty reduction in Thailand, Indonesia and Malaysia had been trumpeted as the showcase of outward-oriented free-market success, just as Japan, South Korea, Taiwan, Singapore and Hong Kong had been in the 1970s and early 1980s. In fact all these but Hong Kong have been showcases of state-led development, but this unpopular truth was largely ignored until 1998. Once the collapse happened, the movement toward a kinder neo-liberalism with a pro-poor face, which had begun in 1990, became a rush.

Writing late in life, the high priest of the free market, economist Friedrich von Hayek (1976: 64) wrote that 'to demand justice from [a free-market] process is clearly absurd, and to single out some people as entitled to a particular share evidently unjust'. Hayek did add the proviso that society has a 'duty of preventing destitution' (which is more than some other neo-liberal philosophers have admitted) thus showing 'a triumph of good sense over doctrine' (Gasper 1986: 164). The idea that there could be a 'redistribution with growth' was first proposed by economist Hans Singer in the report of the ILO employment mission to Kenya in the early 1970s (ILO 1972: 111–14; 365–70). He proposed a confiscatory tax on the top 10 per cent of incomes, to be used for investment in activities that would create employment for the poor. Unsurprisingly, Singer's radical tax proposal was not taken up, but the approach through investment was, and it was built into the theoretical approaches pioneered in Chenery *et al.* (1974). In turn, this led to a strong emphasis in development practice on 'basic (human) needs' in the later 1970s.

The issue of poverty relief had been brought into prominence. It took a sharp dip in the 1980s but even in the starkest days of 'swim or sink' structural adjustment, the need to provide at least some basic support to the very poor never vanished from sight. Several developed countries means-tested or otherwise cut back on the social security systems developed earlier in the twentieth century, but even the harshest (such as Britain under Thatcher) could not abolish them. In Latin America, where the Brazilian military regime had introduced an important welfare support system for farmers and farm labourers in the 1970s, several other countries, including Peru under the harshly neo-liberal Fujimori regime, offered programmes of assistance to the very poor, including the rural poor (Crabtree 2002). On the other hand, many of the world's poor were left to sink into deeper poverty.

In the early 1990s, poverty assessments began to be undertaken on a country-by-country basis, broadened in the later 1990s to become 'participatory poverty appraisals' drawing on the 'participatory rural appraisal' techniques of the 1980s associated particularly with Robert Chambers (1994, 1997). After 1997, the precursors of what became the widely required Poverty Reduction Strategy Papers were first drawn up. These were required first for international debt reduction, and then to qualify for access to development funding in general after 2000. Although said to be 'country-owned', the detailed strategy statements have to prioritize areas that donors are willing to fund, and also affirm the basic conditions of structural adjustment: openness to foreign investment, good financial management, privatization and free trade (Craig and Porter 2006).

Discussing these innovations, Porter and Craig (2004: 411) remarked that they 'do everything possible to create a system of global openness and integration ideally suited to the interests of international finance and capital . . . while depicting the framework as about the interests of the poorest and most marginal people'. These two authors have more than a decade of experience with this process and, while affirming that there is a new turn from frank neo-liberalism, they retain considerable doubts. They note the major erosion of national independence that has taken place in the process of getting resources more effectively to the localized poor. A paradoxical consequence is that international development financiers have now found that they need stronger, 'more capable', national governments – not to obstruct their collective will, but in order to carry out their programmes!

A disturbing interim conclusion

Recalling the discussion of Latin American 'structuralism' at the beginning of this chapter, we now approach a rather depressing interim conclusion. For 'developing countries dependent on exports of primary production' we might read 'family farmers seeking to sell their produce in a world in which that produce is no longer in short supply'. Persistently, their produce prices have weakened while their input costs, and general living costs, have risen. The terms of trade have run steadily against them, even while they have intensified production, whether with machinery and other inputs bought on credit (thus accentuating their dependence on banks or merchants) or by Chayanovian exploitation of their own and their families' labour. Just as ECLA wrote of their countries' problems as structural so, we conclude, are the problems of modern small family farmers. Neo-liberal development specialists and advocates have made conditions harder for farmers in the developing countries, but they were already difficult, and would have remained so even without the doctrinaire new policies that have hit them in the last quarter-century. We defer a final conclusion on this aspect until Chapter 14.

Notes

1 From an early time few have argued that 'growth' is all that 'development' is about. The many definitions of development place emphasis on social change, initially toward the model of 'modernization' or 'westernization', and latterly toward a range of goals which include poverty reduction, education and literacy, and such vaguer notions as the embedding of human rights, livelihood security and 'empowerment'. Nonetheless, economic growth is still generally regarded as the precondition for the wider prosperity that underlies all these desirables. At random, take for example an anonymous reviewer in *The Economist* (29 April 2006: 82), who wrote that in this decade 'economic growth is pulling millions out of poverty'. Yet it is leaving other millions still there.

2 There is no agreed path through this literature, neither its general arguments nor the Latin American branch emphasized in this chapter. Everything is subject to dissent. Interpretations even of events differed strongly at the time and continue to differ. In this book it would be unwise, if not impossible, to suggest particular uncontroversial sources. For ourselves, we cite Warren (1973, 1979) as one of the best sources on the early development period proper, up to the 1970s. Warren was a convinced follower of Marx, but from this point of view he evaluated most of what had happened under post-1945 capitalism as not only successful but even as promising for the future.

3 This committee was formed under the 'Alliance for Progress' instituted at Punta del Este in Uruguay in 1961, a conference convened by the USA under the Kennedy regime after its disastrous attempt to extinguish the Cuban revolution by proxy, and with great concern over the possibility of further peasant-based revolutions in the region. The Alliance for Progress functioned for about a decade, dying away during the Nixon regime. It had generated considerable opposition in the US Congress, as well as among Latin American oligarchs and military officers.

4 In relation to Latin America we use the term campesino, introduced in the Introduction, rather than 'peasants' as in most of the literature.

5 Boards that are legally granted control over the marketing or purchasing of agricultural produce were first introduced in the 1920s and 1930s. Dairying and meat production were among the first industries to be organized in New Zealand in the 1920s as a response to the effective voluntary organization of cooperatives in Denmark since the 1880s. They especially dominated the British market. They employed single-desk selling, price pooling, revenue pooling and preferential financing especially of inputs to achieve higher export prices and hence higher producer prices (Barrett and Mutambatsere 2006). During the crisis years of the 1930s, marketing boards were set up covering a wider range of products including wheat in Canada and Australia, and then in internal trade as well in Britain, where they were felt necessary to rationalize a system characterized by 'an ever-lengthening chain of middlemen' (Stamp and Beaver 1941: 156).

6 One consequence was the proliferating regional and local unionization of Brazilian agriculture, which remains a strong feature of rural organization in the country to this day.

7 In Australia, where the same technical improvements were profitably adopted, a major modern increase in beef exports is explained by many as the product of global improvements in marketing conditions as a consequence of the limited advance toward free trade in the 1990s negotiations that set up the WTO. Pritchard (2006) reminded that a set of special and unrepeatable conditions was also involved, having much to do with the international impact of diseases from which Australia, like New Zealand, has remained free, and also with changes in dietary preferences in Japan and Korea, the leading markets for both countries. Now, with the next round of international trade liberalization seemingly stalled indefinitely after three failed WTO meetings, and with the emergence of a reinvigorated Brazilian agriculture as a major supplier to the world market, the prospects look less promising.

8 New entry to the system was mainly among farmers undertaking small-scale collecting and retailing. Although personnel had changed as the rural storekeepers were replaced by specialized traders, the chains of credit remained much the same as in the 1960s (Barrett 1997). In the Antsirabe region the small collectors still operated mainly in the villages and the assembling collectors in the country towns. To this level, transport was still largely by ox-carts, and small millers were the main buyers. Ox-carts were often intercepted and their loads bought by wholesalers on their way into the market towns.

9 Farmers as landscape custodians

Environmentalism, land degradation and pollution

Contradictions: Polanyi's 'double movement'

During the first two decades of neo-liberalism (1980–98), the median per capita income of the developing countries stagnated at a growth rate of 0 per cent (Easterly 2001). The arithmetical average was pushed up by China, Southeast Asia and India, but there was also a slowdown of economic expansion in the wealthier member countries of the Organization for Economic Cooperation and Development (OECD). Failing to find any obvious answer in a comprehensive statistical exercise and, as a good World Bank economist not about to blame the free market, Easterly suggested that the gap between rich and poor, which narrowed in the 1960–80 period, had returned to its historical – perhaps he meant structural – pattern. In addition, inequalities in many countries have widened in the same two recent decades and, as we saw in Chapter 8, the incomes of many farmers who had survived the crisis years did not improve.

Yet during the same last quarter-century, farmers – especially those in the developed countries – have been charged with an additional set of responsibilities. They arise largely from wide public recognition of the environmental damage wreaked since the 1950s by modern mechanized and chemicalized agriculture. Developed country farmers are far fewer today than they were in 1950, but they still manage the larger part of the lived-in and worked-over landscape. They are expected to preserve this land from degradation, keep it in sustainable productive condition, protect its biodiversity, prevent pollutants arising from farming and its livestock from fouling public water supplies, keep their livestock under what are now regarded as humane conditions and to maintain a farmed landscape that their fellow citizens find attractive. In the North Atlantic countries, the rise of these concerns facilitated major changes in the nature and structure of subsidies, briefly introduced in Chapter 6; looking after the environment has been, in large measure, compensated in ways further discussed below and in Chapter 10. Elsewhere in the developed world there has been regulation with little compensation, and in the developing countries the new concerns are still only very patchily addressed.

There is some irony in the fact that deregulation of the market has been accompanied by a strong growth in regulation of environmental management. There is more than something of Karl Polanyi's (1944) 'double movement' in this. In creating a market economy, land – like labour – must be treated as a commodity, yet land is not produced in order to be dealt with in this way. In the nineteenth century, the destructive effects of the market economy were most obviously inflicted on society and a 'deep-seated movement sprang into being to resist the pernicious effects of a market-controlled economy' (Polanyi 1944: 76). In the twentieth century, many of the protective devices created for people in the previous century remained intact and were even strengthened in the developed countries until the 1980s. Meantime, and especially after 1950, it was the environment – Polanyi's 'land' – that was now ravaged in a manner 'awful beyond description' (1944: 76). The new poverty reduction policies that we discussed in Chapter 8 are described by Craig and Porter (2006: 14) as a somewhat shallow deployment of Polanyi's 'double movement'. The massive growth of environmental concern since the 1960s is, we suggest, a more profound movement in protection of the ultimate basis of livelihood.

There is a further irony. For farmers to accept responsibility for environmental management means a major limitation to their property rights, so sacrosanct to neo-liberal thinkers. The environment, and many of its elements more specifically, are regarded more and more as public goods. They remain so regarded, whether on public or on private land. Environmentalists, and a large public now well educated by them, see farmers as custodians of these goods, not as unimpeded owners of all that is on their land. Subsidization of farmers for the cost of managing the environment, whether to recompense production losses or to help pay for capital improvements required, is far from universal and has grown up only slowly. Moreover, it is often provided only or most bountifully to the larger farms which generate the most pollution, so that environmental management simply imposed by the authorities can become an additional way of squeezing the smaller farmers.

There was a good deal of environmental protection in practices current before modern times. In Europe, and in other regions as well, a web of local regulations had operated for centuries, with or without codification in by-laws or in contracts between landlords and tenants. The English landscape of fields and woodlands was shaped by such a web of practice that goes back into medieval and even earlier times (Rackham 1986). The medieval manorial system embodied regulation for preservation of specific land uses. Later, eighteenth-century tenants needed to keep their land in the condition required by the landowner if they sought renewal of their leases (Middleton 1798). In regard to water, farmers have been for a long time discouraged from polluting the water which urban people wished to drink, an extension of widespread rural practices under which human and animal wastes are so far as possible managed so as to enter rivers downstream of the stretches used for drinking and washing. Very widely, farmers and rural communities as a whole take pleasure in well-managed environments, and in the past much regulation has been unwritten practice, or sanctioned by social custom or religious beliefs. Conserved areas in many lands were protected by being the known abode of spirits, and such mystical sanctions often survived for many generations after conversion of populations to one or other of the major religions. No higher authority was usually involved.

While this description remains valid for limited parts of the developing world, neither mystical beliefs nor local regulation have ever been able to stand in the way of the forces of 'progress' for long, or of the rising demand of growing numbers of people. The outstanding case is China, where love of the landscape is expressed in the words of many centuries of poetry, but where unremitting pressure to control both the character and processes of the natural environment for human ends has created, ultimately, one of the world's heaviest loads of rural land degradation and pollution (Elvin 1973, 2004; Smil 1984). Regulation and incentive conducive to better environmental management by farmers has now reached many parts of the developing world, China included, but first there was a need for a state of crisis to be perceived, for its dimensions to be explored by science, and for scientific findings to become part of the awareness of policy makers and the thinking public – and of at least a proportion of the land managers themselves. This process began seriously only in the late nineteenth century, and it developed greatly in the mid-twentieth century. It began in the developed countries, and we will explore it in this context.

The rising power of the environmental movement

The public of the world learned about pollution in ancient times, and were first encouraged to worry about soil erosion in the nineteenth century. In the 1940s and 1950s such public alarm as was not generated by actual or potential war had increasingly to do with the rapid growth of population, crowding finite resources of land and threatening the next generation with a 'hungry future'. International bodies emerged to influence policy on environmental management. The United Nations Educational Scientific, and Cultural Organization (UNESCO) and its offshoot the International Union for the Conservation of Nature and Natural Resources (IUCN) were formed before 1950, and powerful non-government organizations such as The Nature Conservancy and the World Wildlife Fund began work in the 1950s and 1960s.[1] The adverse effects of chemicalization on the soil were reviewed at a very early date by Balfour (1943), but little notice was taken of such warnings until the 1970s, when the consequences in terms of reduced organic matter and weaker soil structure began to be felt. Before the effect on the soil, or the opening of the landscape for machinery, became matters for concern, rising worries about environmental pollution received a major boost from a single book by biologist Rachel Carson (1962). Her book, a bombshell about DDT and other chemical pesticides, was dropped into a world that had been growing complacent that chemicals could solve its problems about insects and insect-borne disease, as well as those of agricultural production.

Conservation

There was also conservation. The notion of reservation was applied in the nineteenth century to forests that might be valuable for long-term timber production, and because of supposed effects on rainfall. Conservation for the protection of wildlife arose from the hunting reserves created by medieval kings and early modern aristocrats. Wildlife reservations were first more widely imposed in colonial regions, especially in eastern Africa. In North America, the modern beginning was with national parks from the 1870s onward. Elite groups of scientists and some leading public figures were involved in these activities, but in the early twentieth century the model of national parks began to extend, especially in the USA, to the notion of 'wilderness' as the property of all, even if only a small minority of the public could be allowed to enjoy it.[2] The major expansion and diversification of the environmental movement after the mid-twentieth century demanded a much wider base of public support in the developed countries, and what can only be described as very skilful propaganda was conducted in order to achieve this. It has cost a lot of money, a good deal of it public money. National, international and private bodies were all involved.

UNESCO launched its Man and the Biosphere Programme (MAB) in 1971, and in the same year the most aggressive of the international environmental NGOs, Greenpeace, began its activist work. During the whole period 1960–2007 new environmental concerns arose one after another, only a few of them replacing discredited earlier issues, and for the most part cumulatively adding to the agenda. Words were a major product of the new wave of interest, among the most influential on policy being IUCN's *World Conservation Strategy* (1980) and the distilled wisdom of the World Commission on Environment and Development (1987). These were backed up by a great wave of 'ecodoom' writing, the very exaggeration of which gave it greater effect in forming public opinion. Major world conferences were held, in Stockholm in 1972, Rio de Janeiro in 1992 and Johannesburg in 2002.[3] The first of these set up the United Nations Environment Programme (UNEP), and the second led to some important international agreements. Between and around these events, a multitude of books, newspaper articles and, perhaps above all, television programmes, created a wholly new degree of environmental awareness that has now become almost truly world-wide. It has reached many farmers as well as the larger non-farming public. We focus on three issues, to demonstrate how farmers in different world regions have been impacted and have responded in different ways. These are land degradation and agricultural pollution in this chapter. Biodiversity issues are taken up in Chapter 10.

Land degradation

As a topic of concern, land degradation substantially preceded the modern era. In western Europe, there had been periods of significant soil erosion in the fourteenth and eighteenth centuries (Blaikie and Brookfield 1987) and, mainly because of what happened in the USA in the 1930s, it was the central environmental issue in the 1930s and 1940s. Soil erosion has long been seen as a major problem in southern Europe but not in the north until recent years, when it has again arisen under the impact of modern farming technology (Boardman and Poesen 2006). In the USA, on the other hand, soil erosion has consistently been of concern since the 1930s. A good deal of the 1930s assistance to farmers in the USA was dressed up in conservationist clothing, although its real purpose was income support (Batie 1985).

American farming had been fairly casual in its land management practices for a long time.[4] In the 1970s, the short-lived boom described in Chapter 6 led to the new cultivation of as much land as all the wheatlands of Canada, using machinery larger and heavier than any used before, and there was a substantial new surge of soil erosion (Blaikie and Brookfield 1987). A significant proportion of the control works created during the New Deal period and up to the mid-1950s had been allowed to fall into disuse, or had been ploughed over in larger fields. Pavelis (1983) described this as consumption of capital due to pressure to produce. The result of rising conservation concern was the set of conditions incorporated first in the Soil and Water Resources Conservation Act of 1977, and then in the more far-reaching US farm bill of 1985, described in Chapter 6. Penalties were applied for the new cultivation of highly erodible land without conservation measures, or of unused wetland. These regulations apart, American farmers have been turned from wasteful ways by subsidy.

The 1985 bill incorporated a voluntary 'conservation reserve program' (CRP) that went far beyond anything in previous legislation. It invited farmers to apply for funding to set aside parts of their land for ten-year periods for the purposes of erosion control. For this they were and are paid an agreed rent together with some support for conversion and management costs. Selection from among applicants has always depended partly on costs, using something like a reverse auction system in regard to the proposed rents, but as the scheme was continued under successive farm bills, an 'environmental benefits index' was elaborated to give better weight to the several objectives.[5] CRP land could be re-contracted for further periods, subject often to some re-seeding or other management provisions, or it could be rotated with other land, and from 2004 early renewal of contracts became possible. By the turn of the century, 180,000 sq. km of land were enrolled in the CRP, distributed very unevenly across the country. Then the 2002 farm bill added a further important support measure for conservation works on utilized land, seeking to restore some of the conservation capital lost between 1950 and the 1980s.

Soil conservation in the developing countries

Farmers and pastoralists in developing countries of Africa and at least southern Asia heard about land degradation long before their fellows in the developed countries. From the late nineteenth century they were being told of the ills of 'over-grazing' with their cattle, and abused for the 'primitive' nature of their farming systems. By the 1930s, when American farmers were first being assisted in the construction of conservation works, African pastoralists and mixed farmers were being instructed to cull their herds of cattle, and commanded (with little success) to give up shifting cultivation and to construct terraces on their cultivated hillsides. The spectre of 'desertification' was first raised in these years and their 'ignorant' ways were given a prime role in the wastage of land first discussed on a world-wide basis by Jacks and Whyte (1939). The pressures continued and intensified in the last days of colonial rule after World War II, and generated a great deal of hostility leading, in Kenya and a few other places, to active rebellion. Farmers in India were mainly being ordered to cease cutting into the remaining forests. In South America, by contrast, the subject of land degradation was barely raised before the 1950s except by a few academic observers.

For Africa, at least, the early emphasis on soil erosion doubtless arose from the prominent gullies that scar extensive areas. Although Stocking's (1978, 1996) finding, that the gullies mainly carry away the soil washed into them from the interfluves, has been widely confirmed, the emphasis on erosion continues. There are good reasons. Most soil organic matter and available mineral nutrients, especially nitrogen and phosphorous, are in the surface horizons. Their removal by water or wind leads to sharp losses in soil capability, contributing greatly to the declining fertility which is a wider complaint than erosion itself in tropical and sub-tropical countries. Recent emphasis is more on sheet-wash, rills and soil creep than on the gullies (Stocking and Murnaghan 2003). Most of the indigenous management tools informatively presented by Reij *et al.* (1996) are concerned with such less visible processes. Intervention projects, which have widely succeeded the command-and-control approaches of the colonial period, tend to like the larger engineering works such as terraces, cut-off drains and infiltration ditches. Command-and-control has not ceased; for example a badly eroded area in central Tanzania, on which British experts had worked rather ineffectually for 50 years, was forcibly completely destocked in 1979 during the time when Tanzania's dispersed rural population was, with equal determination, being collected into large villages for socialist development. This 'Kondoa Eroded Area' was still without livestock in 1995, when there had been some significant recovery of ground cover (Östberg 1995).

Command-and-control has sometimes been effective where the methods imposed were suitable to deal with the problems, but where large-scale soil engineering works were constructed, maintenance has been a problem and cut-back terraces have led to the cultivation of subsoils. The indigenously developed 'thrown-up' (*fanya juu*) terraces do not suffer from this problem. Once the pressures were relaxed, farmers have frequently neglected the 'modern' works, including those constructed for them by machinery or compulsory labour. The same is true of a great many of the 'voluntary' projects that succeeded colonial direction after countries became independent. Often financed by national aid agencies, projects of the 1960–80 period suffered a high proportion of failures, leading to important changes in the direction of greater participation in planning by the farmers, following the urging of Chambers *et al.* (1989) and others. Or at least it led to changes in rhetoric, if not always in practice.

A mark of change is a relatively successful intervention in Burkina Faso. During the 1980s, dissatisfaction with the many failures of top-down intervention, and decisions to take more positive steps following the famine relief of the 1970s drought in the Sahel, led to a new approach in francophone Africa termed *Gestion des Terroirs Villageois* (GTV). It involved close village participation at all stages including the selection of technology. The aim is to create – or enhance – the ability of communities to manage their own land (*terroirs*) through committees formed for the purpose, with scientific and logistic aid. In the uncertain rainfall conditions of the Sahel it was of particular importance to improve the conservation and harvesting

---

BOX 9.1 PEOPLE AND A SOIL CONSERVATION PROJECT IN TANZANIA

Östberg (1995) provides revealing detail on people–project relations in soil conservation matters. Outside the Kondoa Eroded Area, Burunge farmers have individualistic farming practices, but individually held land becomes a common grazing resource during the long dry season when no crops are in the ground. A large Netherlands-funded soil conservation project in the Kondoa area started work among the Burunge in 1994. They came with an effective technology for water retention, by means of an 'infiltration ditch' with cross-ridges, fed by permanent cultivation ridges designed differently from normal Burunge practice. Some of the better-off farmers welcomed the project, expecting to have work done for them and to gain prestige. The project waited for individual invitations. These did not come, because the larger farmers waited for prestigious town-based experts, while the others waited to be employed and to receive money and food. Eventually one medium-scale farmer did invite project intervention.

Work on his new ditches began promptly, but they were trampled by cattle going between their pastures and the village. This led to dispute between the farmer and the cattle owners, and other farmers became anxious that the infiltration ditches would lead to effective land privatization, to exclusion of cattle and to orders from government to do compulsorily what had been presented initially as voluntary. They and others had experience to help generate this fear. They felt that their existing, annually constructed, cultivation ridges would do the job well enough without leading to these problems, and the inevitable disputes between neighbours. Beyond that, they knew that promised yield increases had not eventuated in the nearby eroded areas, where soils made largely of downwashed sands retained little water. Finally, they believed that water derived from upslope forests would bring 'strength', which they knew already how to manage. The project had arrived with a ready-made solution, one which failed to take account either of existing seasonal rights or existing environmental understanding, which was contemptuously disregarded. To Burunge farmers, it was yet another demonstration of the coercive and unequal relationship between the global centre and peripheral farmers.

---

of water. The methods proposed to village committees in central Burkina Faso included small dams, cultivation pits and, in particular, stone lines along the contour (*diguettes*), a hybrid of indigenous and scientifically promoted practice. On the gentle slopes of the central plateau of the country, the purpose was to trap the flow from short, heavy rains, and the topsoil and nutrients carried with it (Batterbury 1998).

Locally shared management and good advance preparation led to considerable enthusiasm, although in the dry season (when the stone lines had to be laid) many men were away at work and almost all the heavy labour of collecting the stones, piling them onto trucks and laying them along surveyed lines was done by women. By the end of the century there were 240 Burkinabe communities participating in GTV projects, although the external financial aid was likely soon to end. Batterbury, who worked with a GTV project while a graduate student in the early 1990s, revisited the same villages in 2001. He found, as did others, that the *diguettes* had indeed worked as planned and that there had been an improvement in millet yields, but that many local problems of unequal access to the benefits remained unsolved. Importantly, his principal village had received a new school and a neighbouring community had acquired a deep well.

The well-publicized evidence of community participation had generated 'symbolic capital', which was in these two cases transformed into valuable concrete (Batterbury 2005). Indeed, village participation in a land degradation project had been viewed by the people long before this in terms of what aid it would bring, rather than for its conservation value alone (Batterbury 1998). Land degradation remains of greater concern to international funders, and to governments which wish to gain access to their support, than it does to most of the developing country farmers themselves. Meantime, it has supported a large world-wide body of consultants. As we shall see in Chapter 10, the same is true of the newer international involvement in biodiversity conservation which, in relation to the developing countries, became a matter of overriding international concern.

Agricultural pollution: Europe and North America

In Europe, the first substantive issue was pollution of water. Management of urban and industrial pollution goes back a long way in human history, but it was not until the 1960s that pollution generated by agricultural practices became a matter of wide concern. Nitrogen and to a lesser extent phosphorous were being applied in large quantities to crops; high nitrogen loads in animal feedstuffs were further adding to an excess of nitrogen in the soil which then drained or leached into water. Contamination of drinking water by nitrogen and pesticide residues, as well as by bacteria, led to the imposition of mandatory standards for public supplies in many countries by the early 1970s. The cost of purification was high, and this was by no means all the problem.

One consequence of developed country subsidies to their farmers, and rising markets for meat, was the intensification of livestock production, including large feedlots for the fattening of cattle and – with greatest environmental consequence – more malodorous pigs. Battery production of eggs and chickens also expanded greatly. Further, the crops grown for livestock fattening were heavily fertilized. As production intensified, the old straw-bedding system of animal keeping, mentioned in Chapter 2, increasingly gave way to slurry systems. The sheds had a channel that wastes were washed down, to be stored in ponds or tanks, from which the slurry was later sprayed onto the land. While the bedding system, which created stackable manure, allowed the liquid pollutants to escape, the small scale of operation dispersed the problem. With the growth in scale, and increasing use of the labour-saving slurry system, the scope for pollutants to enter the ecosystem was hugely increased.

Whole regions in which these activities were most intensively pursued became severely polluted. High concentrations of nitrates and especially phosphorus in water led to eutrophication of water bodies, including rivers and coastal waters, prompting problems for fisheries and also tourism. In the USA the seriously affected areas included Iowa, North Carolina and the southern catchment of Chesapeake Bay. In Europe they were even more widespread, including most of the Netherlands, large parts of Belgium and Denmark, Lombardy in Italy, Galicia in Spain, the west of England and Bretagne (Brittany) together with adjacent parts of Normandy in France. In all these areas, livestock pollution was added to the nitrogen supplied by mineral fertilizers in quantities far larger than could be used by the crops or grasses.

Nitrates were the first and for a long time the only issue in most regions because of their effect on drinking water. In the regions of intensive livestock production, both surface and groundwater regularly exceeded the tolerance level of 50 mg/l of nitrate set up by the WHO in 1975, and accepted as standard in Europe. The American Clean Water Act of 1971 preceded this, but it was implemented only at state and municipal levels, and it took a further 25 years before an integrated strategy to cope with agricultural pollution was undertaken.[6] During this same long period, livestock farming became less and less a family or small farm operation, and more and more industrial at the final, fattening stage. To a lesser degree, the same was happening in Europe. The management of agricultural pollution under these changing structural conditions presents a revealing story, which we tell in some detail because of what it reveals about policy and behaviour.

Foot-dragging in Europe

The biggest problems lay in the handling, storage and use of livestock manure. It was historically applied to the land in the spring, when it is most useful to plants, but with the increase in quantity it was applied at less useful times, even in the winter when there was minimal uptake by plants. It became necessary to restrict the amount applied to land, to regulate the season of application, and to improve storage conditions to reduce leaching into the soil, spillage into open water and distribution by the wind as dust. Especially with the more noxious droppings of pigs and chicken, neighbouring householders began to complain vigorously about the odour. When and where adopted, all public regulation has addressed these problems, by limiting the amount applied to the land, prohibiting it in the winter and sometimes also the autumn, requiring adequate confined storage and establishing minimum separation distances between the operation sites and the land on which manure is spread, and neighbouring houses, public buildings, parks, roads, streams and sources of drinking water and water used for recreation. The issue of nitrate pollution of water provided the means for addressing a whole related set of issues, but in ways that cost the livestock producers money and aroused some powerful opposition.

The strongest and earliest action was taken in the Netherlands and Denmark, the former having the highest livestock density of any country in the world together with a high water table over most of its territory (Debailleul 2002; Fraters *et al*. 2004). Both countries acted in the early 1980s to impose management standards on the handling and use of livestock manure. In 1991, after several years of discussion, the EU issued its landmark Nitrate Directive, not a law but a set of guidelines with which all member countries were required to comply. Its aim was reduction of nitrate content in water to not more than 50 mg/l, and for this purpose it identified 170 kg/ha as the maximum fertilizer or manure application to agricultural land. In dealing with livestock manure only, this implied a maximum carrying capacity of only two livestock units per hectare.[7] The directive, strengthened in 2000, called on all countries to report on conditions, to identify land vulnerable to nitrate pollution, develop action plans that would reduce nitrogen and (later) phosphorous pollution, to establish mandatory codes of good farming practice and to aim rapidly toward the WHO standard for all their potable water supplies. Where 'nitrate-vulnerable zones' were declared, controls on farm practices were mandatory within them; elsewhere they were voluntary.

The response was extremely mixed, with most of southern Europe virtually ignoring the directive in favour of pressing forward with production-oriented support measures. Several countries found the determination of 'nitrate-vulnerable zones' difficult technically, and many farmers disputed their inclusion within them. Discussion went on for years. In northern Europe by 2002, there was some improvement in the quality of surface and shallow water. It would inevitably take 40 years or more before major improvement could be expected in deep ground-water, in such environments as the loess regions of Europe and the chalk regions of Britain. There were big gaps in actual performance. The fact that in the worst affected areas, with the highest livestock densities, the only long-term solution involves a reduction in livestock numbers and densities was understood early, but such action continued to be resisted by the producers and the economic and political interests that support them.

The scale of the problem is well illustrated by the Netherlands. On the basis of a mineral accounting system introduced in 1998 it was shown by Fraters *et al*. (2002) that nitrogen input to surface soil averaged 403 kg/ha, of which 195 kg/ha were by manure, 155 kg/ha by mineral fertilizer and most of the balance by atmospheric deposition. The average nitrogen surplus over the whole country was an alarming 190 kg/ha. From 2002, livestock farms adding more than 170 kg/ha for arable land, and 250 kg/ha for grassland, have been required either to sell the surplus manure to other less intensive farms, or reduce their livestock numbers, but this has been hard to enforce. Although notable improvements had been achieved since 1994, it was clearly going to take a long time to achieve the EU standards in the most intensive livestock farming regions. Progress has been slow everywhere, even in France which in 1993 embarked on a national programme to subsidize livestock producers to invest in facilities for better management and disposal of pollutants – at first outside the Common Agricultural Policy, and only after 2004 within it and using a proportion of EU funds.

BOX 9.2 A NITRATE-VULNERABLE ZONE IN SCOTLAND

A case in Fife, Scotland, is recounted by McGregor and Warren (2006). Here, a nitrate-vulnerable zone was declared only in 2003, to the great consternation of arable and livestock farmers who had believed that their practices were causing no problems. The costs faced by the dairy farmers for investments in manure and slurry storage were considerable, despite a potential 40 per cent subsidy. Arable farmers, who had been exchanging straw for manure with livestock farmers, were uncertain if the exchange could continue. Nutrient budgeting, which could answer many of their questions, was an innovation outside their experience. Meanwhile, these inland farmers were unable to fully perceive the linkage between their practices and the condition of the estuary of their river, the cause of the declaration. The establishment of buffer strips near streams on which no fertilizer or pesticide could be sprayed would lead to costs in terms of lost production and weed infestation. Soil testing, as required in the plan, was seen by many as not necessary, a bureaucratic imposition. Lack of information and a distrust of bureaucracy dominated farmers' reactions. The Fife farmers were not alone in this.

Bretagne, in western France, as recently as the 1960s was still a land of small farms that were mainly producing livestock and vegetables in a landscape of hedges and small woods. Since the 1970s it has been peppered with large livestock feeding operations for cattle, poultry and pigs. By the end of the century it had become France's leading region for large-scale livestock farming, with 55 per cent of France's pork production. With only shallow aquifers in weathered metamorphics, run-off to rivers and the sea has been rapid. Since the 1980s problems of excess nitrate in water, and eutrophication in rivers and estuaries have been compounded by heavy seaweed growth on shorelines. The smell of decaying seaweed has depressed parts of Bretagne's large tourist industry (Bureau 2003).

With only slow uptake of the national programme, and a lack of enforcement, a civil society organization, Eaux et Rivières de Bretagne, complained to the European Court of Justice which, in 2001, issued a condemnation (a warning) to the French Government for violation of the 1975 Clean Water Directive in regard to Bretagne (Begat 2001). By 2003, 13 of the 15 member countries of the then-EU, excepting only Sweden and Denmark, had been cited for slow progress under the 1991 Nitrate Directive. New and more precise legislation followed, but the foot-dragging has continued. Where money was made available for remedial works, it was far less than the lost production subsidies. Thus even in France, where legislation is strong and monitoring capability is excellent, and where modest funding for capital improvements was authorised well ahead of EU agreement, the real problem of localized livestock overpopulation and consequent pollution continues to be avoided. The restructuring involved is just too much for either national government or regional authorities, let alone the farmers. Quite often farmers disbelieve assertions that nitrates are a problem. The conflict of objectives is rather starkly revealed in Spain, which entered the EU in 1986 with an agriculture that, notwithstanding major investments in the preceding 25 years, was seen as still in serious need of modernization. Public and private investment continued to address the need to improve water supplies for irrigation, improve rural infrastructure and modernize farming methods. At the time of accession, government was attempting to restrain the heavy drawing down of aquifers which was leading to saline intrusion in coastal areas, but their actions had virtually no effect on continued private expansion (Izcara-Palacios 1998). Nowhere was nitrate pollution perceived as a problem so that the EU directive of 1991 was at first ignored by both national and provincial governments, then ignored by irrigation farmers who were more concerned with the competitiveness of their production in the enlarged market.

CAFOs and local initiative in the USA

Environmental topics in Europe, even if directly related to agriculture, still rest largely in the hands of the member nations. In the USA they remain to a large degree in the hands of the constituent states. From the 1970s, another dimension of the problem arose from the emergence of very large concentrated animal feeding operations (CAFOs), which were slower to emerge in Europe. If we were writing of a war between capitalist farmers and family farmers, then the CAFO would surely be one of capitalism's most powerful weapons. Its facility for vertical integration with meat packing and processing, the ability to automate feeding, and its low labour requirement in relation to number of animals, all give it great advantages in the cutting of costs. Control over slaughterhouse facilities, from which many CAFOs evolved, gives the capitalist enterprise the means to contract individual farmers to provide young livestock that can then be fattened. This has short-term benefits, but when livestock prices decline because of the large production generated by the industrialized system, or for other reasons, many smaller livestock farmers are unable to compete and either enter wholly into contract with the CAFOs, or go out of business.

The CAFOs have generated sustained hostility, not only from smaller farmers but more effectively from the rural non-farm population that is affected by the large-scale pollution generated. They lobbied strongly for controls. The state of Iowa already had wastewater and zoning laws in place in the 1980s and these were sufficient to block plans for a large new CAFO operation in the state. The company concerned was able to move to neighbouring Missouri, where a severe downturn following the depression of the early 1980s (Chapter 6) led to a welcoming political environment for large new enterprises (Constance *et al.* 2003). The welcome went so far as to create a special exemption to a state law of 1975 preventing the entry of new corporate farms, enacted with the aim of protecting family farmers. A very large vertically integrated operation was set up, extending from breeding piglets to processing and marketing. Mexican rather than local workers were engaged. Two other companies followed, and by 2000 all three were in the hands of America's two largest pork and pork-product companies.

In 1995, nine spills from clay-lined slurry pits caused extensive fish kills in the rivers, leading to new state legislation to require more secure waste management. There were even worse spills in North Carolina, and large fines under the Clean Water Act were imposed. Belatedly, it would seem, the national Environment Protection Agency became involved in legal actions. By 2002, stricter controls over CAFO operations prompted the companies to give up further expansion in Missouri, and shift their growth to Texas and Oklahoma. By 2001, there were an estimated 6,000 CAFOs in the whole of the USA (Claasen *et al*. 2001), and attention under the Clean Water Act of 1971 had been strongly drawn to them. Many lacked adequate discharge control, and enforcement proved difficult. In 2002, the national Environmental Quality Incentives Program (EQIP), enacted in 1996, was substantially enlarged to provide more funding for livestock producers, and especially the CAFOs, to improve animal waste management with cost-share funding up to 75 per cent. EQIP funding was planned to increase very substantially up to 2007 but, as with all voluntary programmes in the USA, available funding has been restricted by the growing demands of defence and security in recent years.

In Canada, more than in the USA, local government has principal authority over pollution issues. In southwestern Ontario, where livestock farming has become the most profitable enterprise in recent years, amalgamation of local government areas in 2001 placed some farming districts in a new municipality with an urban majority population, itself sensitized to pollution problems by a nearby outbreak of livestock related *E. Coli* infection in 2000. In 2001, farmers' principal concern, other than poor received prices and high input costs, was the probability that much more severe restrictions on disposal of livestock wastes would be imposed by local government, escalating their costs (Smithers *et al*. 2005). Since they already had to compete with capitalist enterprises, though of a smaller scale than in America, this was worrying to well-established family farmers.

Problems in a country in transition: Poland

To conclude this discussion, it will be useful to return to Europe where the accession of new countries has required adoption of policies developed earlier in the 15-member EU. Poland offers a case in point. Even in law, there was no regulation of agricultural pollution before 2000. Most of the country did not experience collectivization and the average farm size is 7.7 ha. Farms between 5 and 10 ha, almost all family farms, have half the total livestock in the country, and 63 per cent of production (Karaczun *et al*. 2003). Although Polish agricultural authorities had noted agricultural pollution in the 1980s, environmental issues were – as in all the east European countries and Russia – considered mainly as urban and industrial problems. It was not until Poland began to negotiate for its 2004 entry into the European Union that the need to adopt the Nitrate Directive was presented to a government and population that did not see agricultural pollution as a problem (Gorton *et al*. 2005). The normal practice with livestock manure was to stack it on the ground, allowing liquid waste to run away. Together with the input from pit toilets and septic tanks, this had the consequence that in one large county northeast of Warsaw with 85,000 cattle and 88,000 pigs at stocking densities less than one livestock unit per hectare, 50 per cent of wells had higher than allowable concentrations of nitrates (Karaczun *et al.* 2003). Limited efforts were made to provide farms with slurry tanks and concrete manure pads before accession date in 2004. Poland has until 2008 to comply with the Nitrate Directive and other environmental directives of the EU, but there was little sense of haste.

There are social implications of the Nitrate Directive in Poland, and in other east European countries that have recently, or soon will, become part of the European Union. Many farms are smaller than 10 ha, and since the capital costs of compliance are scale-neutral to a considerable degree, they fall more heavily on the smaller farms. Such financial support as is provided goes to the larger and more commercial farms, dairy farms in particular. Therefore, write Karaczun *et al.* (2003: 36) 'in both Poland and Lithuania it is expected that the majority of the smaller farmers will never acquire proper manure storage facilities. Due to their uncertain future, they are not targeted to comply.' 'Their uncertain future' means that they are expected to be unable to compete with larger farms, and to go out of business. Yet Karaczun *et al.* (2003) record that in Poland as a whole 75 per cent of farm animals are still kept on the litter system and only 25 per cent on the more polluting slurry system. For small farms with few livestock in fully rural areas, the containment facilities required by the EU are probably superfluous in all but the most sensitive natural conditions. Nonetheless, the directive was incorporated into the Polish fertilizer law of 2000. One consequence of pollution control applied in regions of small family farming is therefore likely to be closure of livestock production on small farms. The preference given to larger farms does include many under family operation, and is not a preference to a few foreign CAFOs as some observers have claimed.[8] This should not surprise, because that part of the 'double movement' that is concerned with environmental improvement is not here in obvious conflict with an opposite movement that puts the 'inefficient' small farm in growing jeopardy. The world can go on demanding that more care be taken of its environment while at the same time getting rougher for the small farmer.

Notes

1 Retaining its acronym, WWF is now the Worldwide Fund for Nature.
2 The notion of 'wilderness' was slow to arrive in Europe, where there is little that could be called true wilderness. It did emerge at the end of the twentieth century in the purchase of profitless farmland by conservationist groups, and sometimes by local or national government, for the specific purpose of creating new wilderness that would protect wildlife, but which at least some of the public might enjoy.
3 The events leading up to the first major conference on environmental issues, in Stockholm, and the course of debate in and around the conference itself, are well described by a Canadian journalist (Rowland 1973). This conference, which had fairly quickly followed the initiation of UNESCO's international project of scientific research, the Man and the Biosphere Programme in 1971, was the origin of the UNEP. It signalled the rapid growth of international concern and was the occasion of the first

specific spread of this concern from the developed countries into the developing. To mark this change, and help ensure its continuity, UNEP was set up with headquarters in Nairobi, Kenya.

4 One of the first modern conservationist writers, Aldo Leopold (1966: 245), wrote to caricature an American farmer of the 1930s period who 'clears the woods off a 75 per cent slope, turns his cows into the clearing, and dumps its rainfall, rocks and soil into the community creek [yet] is still (if otherwise decent) a respected member of society'.

5 The environmental benefits index used to rank proposals covers soil erosion, water quality (surface and ground), wildlife benefits, 'enduring benefits' of the proposed action, air quality benefits and an evaluation of the cost of environmental benefits per dollar expended. The asking rental became critical only within brackets of constant environmental benefits index (Allen and Vandever 2003).

6 In 1997, comprehensive nutrient management plans were made obligatory for industrialized concentrated animal feeding operations (CAFOs) by 2002, and for all the rest of the USA's 450,000 livestock farms by 2009.

7 The definition of a livestock unit, based on cattle, varies considerably from country to country, stock being sometimes evaluated on the basis of weight, sometimes of nitrogen production in manure, sometimes on a more arbitrary basis. Denmark, after twice reducing the equivalence for pigs, finally settled on one sow plus her progeny as the equivalent of one cow (Debailleur 2002).

8 Certain foreign CAFO companies have entered Poland, one buying the controlling share in a formerly state-owned Polish meat company, then financing a front company which could legally buy land as a Polish national. The object was to acquire a stake in the enlarged European market from a base in which land and labour were cheap. Much of the land acquired was that of state farms of the communist period, set up especially in the northern areas that had been part of Germany (East Prussia and Pomerania) until 1945, and in which the large estates of the 'junkers' had until then survived. The company runs CAFO enterprises, fattening huge numbers of pigs, employing little local labour and happy to follow existing national practices of waste management, at least for the time being. Near the new CAFOs, local pollution problems now suddenly appeared on a scale quite outside anyone's experience. But the companies brought export business into a depressed economy, and even though overproduction led to low prices and much distress to farmers, both local and national authorities continued to welcome them. There is much journalism on this topic, but we refrain from citing it because some of it seems more concerned to besmirch the reputation of one American company than to report objectively on the Polish situation.

10 Conservation and growing complexity since the 1980s

In the 1980s a world that a decade earlier feared widespread famine found itself with an abundance of food, if it could have been delivered to all those in need. Famines continued to arise in Africa, but the developed world's agricultural decision makers occupied themselves with problems of surplus production more real than those of the 1930s. This new situation offered scope to deal not only with the old issues discussed in Chapter 9, but also with matters less directly related to human welfare. Biodiversity and conservation of natural resources became central issues for the global environmental movement.

The rise of biodiversity and its conservation

Although biodiversity includes all animals and plants, soil flora and other living creatures in and on the land, until recently the conservationists paid principal attention to vascular plants, animals, fish and birds. More widely than the land degradation and pollution questions reviewed in Chapter 9, biodiversity concerns impact on farmers through a shifting blend of prohibitions, encouragements and incentives. The forces leading to more conservationist forms of agriculture are still gathering strength. They now intersect with larger questions of rural development, and with seemingly contrary neo-liberal trends toward what passes for free trade.

'Fences and guards' conservation

The term 'biodiversity' was first used in 1986 (Wilson 1988), but had been preceded by the substantial creation of nature reserves from which farming and all extractive activity was legally excluded. After 1950, concerns arose about the rapid increase in tropical deforestation for the expansion of agriculture, pastoralism and extraction of timber. Central America and the Amazon basin were the first to gain international attention, although at the time higher rates of deforestation were taking place in Southeast Asia. By the 1970s it was realized that the tropical rain forests contained a high proportion of the world's remaining plant and animal species, and that they were under an unprecedented scale of attack. For as long as it was possible to do so, and even today, blame was placed on the cultivation practices of indigenous farmers rather than on commercial entrepreneurs and state-sponsored development schemes. To most conservationists, the best way to deal with the loss of biodiversity was to reserve representative areas in which the associations of species that constitute ecosystems could be preserved to evolve toward stable 'climax' states, believed in by most ecologists until the 1970s and 1980s. Such 'natural' vegetation and the wildlife it supports were seen as 'pristine wilderness', to be preserved in its equilibrium by excluding farmers. Such protection has extended substantially in the past 25 years. Urged on by some very powerful international conservation organizations, it has become politically desirable to conserve about 10 per cent of the area of each biome in each country.

This target has not been achieved, nor is it likely to be achieved, for a number of reasons. In the first place, it costs money. Only biomes that have popular appeal, outstandingly the tropical rain forest, receive large international funding for total conservation. Similarly, threatened extinction of certain 'flagship species' is an issue that attracts big money, whereas extinction of microbes, insects and most small fauna (with the important exception of birds) does not. In the second place, more farmers and pastoralists are affected as conservation areas expand, and potential or actual conflicts abound. Moreover, there are large scientific doubts about the capability of total protection to achieve its aim, unless the areas are very large. The doubts arise mainly from the dynamic nature of all ecosystems, and from the edge problems of small blocks. Ecology, with its recent emphasis on disturbances as being essential for the creation and maintenance of diversity, emphasizes that only in large and interconnected areas can the patches at different stages of evolution add up to a single dynamic complex. Nonetheless, even now a major extension of total protection is still seen by a considerable number of scientists as the only way to save at least the so-called 'biodiversity hotspots' (e.g. Pimm *et al*. 2001). Biodiversity 'hotspots' are areas that contain a high percentage of all the world species, and which are also severely threatened (Myers 1988; Myers *et al*. 2000). Many have large human populations.

Involving farmers

Established farmers know a lot about the plants and animals on their land. In the early days of the subdiscipline now known as 'ethnobotany', Levi-Strauss (1966) noted that lists compiled for non-literate farming people totalled between 300 and 600 taxa. Subsequent work has confirmed this mean range (Voeks 1998). It is therefore unsurprising that taxonomists recording biodiversity have often relied heavily on farmer informants and assistants. Such knowledge is widespread, and it does not arise from any academic interest in the flora and fauna, but because people find the plants and animals useful or inimical, or just observe them out of interest as members of a common environment. There is an obvious 'contradiction' between farmers' knowledge and management of the 'wild', on the one hand, and the 'fortress conservation' policy of excluding them. Since the early 1990s attention has shifted increasingly to the more conservationist management of occupied areas, and more than half of the world total of declared 'protected areas' as recognized by the IUCN – collectively about the size of Australia, and over 10 per cent of the earth's surface – now consists of 'managed-resource' protected areas, involving farmers and other land users. About two-thirds are in the developing countries (Zimmerer 2006). The probability is that the interference will grow, because both the public and private international environmental organizations regard themselves as having a stewardship role in ensuring biodiversity conservation. They have great power to influence impoverished developing country governments, both through their own substantial funds and by their leverage over much larger general-purpose funding sources. At the present time there is a tendency of opinion back toward 'fences and guards' conservation, since local management is not, it seems, delivering the goods. But cost, as well as other considerations, constrains such a movement.

Happily, our own biodiversity project, People, Land Management and Environmental Change (PLEC),[1] was not concerned with the aggrandizement of conservation territories. We did encounter a number of cases where such aggrandizement had occurred. A forest reserve was declared in 1978 and one Chinese village lost almost half the land allocated to it when the villagers were moved to their present site only seven years earlier. While they continued to use some of the reserved land for a number of years, and to hunt wildlife in it, both these illegal activities were halted by the police in the early 1990s. In Thailand, a minority group community without legal rights to its land had to watch officials of the Royal Forest Department plant fast-growing eucalypts on many of its fields in order to reclaim them as part of the forest reserve. They had no legal means of redress.

Agrobiodiversity

While concern for the tropical forests was still building up, it was realized in a smaller scientific community that the genetic diversity of crop plants, and to a lesser degree farm animals, was threatened. It was being rapidly eroded by standardization and by the replacement of locally bred 'landraces' by germplasm bred on research and experiment stations and either distributed or sold to farmers. Collection and long-term storage of seed, begun in the 1920s, increased massively in the 1960s and 1970s, leading to the assembly of large seed collections in safe storage under controlled conditions of temperature and humidity in seedbanks. By the 1980s the seedbank strategy was being questioned on a number of grounds. Crop plant and other diversity is constantly evolving and the storage of seed freezes this process. It was, therefore, increasingly argued that a great deal of diversity is in farmers' fields, and that it was necessary to conserve the diversity there, *in situ* on farms, with the costly seedbanks playing only a back-up role. Bold new strategies were required. These issues came together in the 1990s and generated wide interest in farmers' methods of seed acquisition (Almekinders and Louwaars 1999; Brush 2000). Agrobiodiversity, involving all crops and livestock and their wild relatives, together with interacting pollinators, symbionts, predators, pests and parasites (Wood and Lenné 1999), is a large and still growing field of knowledge, but the context of farm management has not systematically been explored except by ourselves and colleagues in PLEC. We termed the whole larger field of study 'agrodiversity' (Brookfield 2001; Brookfield *et al.* 2002).

Taking this wider view, our 12-country project set out with the hypothesis that agricultural management employing diversity strategies can sustain and even enhance biodiversity. This view gains support in Europe where until the development of modern technology in the 1950s a thousand years of agriculture had the effect of creating a dynamic mosaic of habitat or ecotope patches that enhanced species diversity, structural and functional diversity, and probably genetic diversity as well (Waldhardt *et al.* 2003). Our purpose was to study farmers' management and its effects, and to encourage good and innovative practices developed by the farmers themselves. Often these included simultaneous management of agricultural, agro-forestry and forestry resources in a single field, a management style common to smallholder enterprises in much of the tropics. Despite their pervasiveness, these approaches are only rarely mentioned in the literature, and appear to be invisible to most researchers. Many farmers manage annual crops in their fields for harvest in a few months, while also tending interspersed tree seedlings that will grow and produce fruits or be cut in thirty years or so for timber. The tree seedlings may either be spontaneous volunteers or transplanted from neighbouring forests or gardens. While the crops are planted, weeded and harvested, the slower-growing trees may receive little more than a cursory weeding and an occasional pruning. The knowledge that local farmers have of the growth characteristics of many organisms and their combinations, as well as of the specific capabilities and limitations of each corner of their fields, can make such complex management profitable (Brookfield and Padoch 2007).

Though not specifically concerned with the *in situ* on-farm conservation of locally bred landraces, PLEC none the less formed a part of that movement (Brush 2000). Seeking the best and most diversity-conscious farmers in each of our more than 20 demonstration site areas we encouraged them to instruct and advise others in the field and on their own farms. Except in a few regions this was not difficult once a larger population accepted our 'expert farmers' as people who really had something to show and teach them. Unable to provide significant direct reward, we encouraged innovations that created added value from biodiversity in the field and in the wider farming environment. PLEC left a substantial publication record, including many papers and four books (Brookfield *et al.* 2002, 2003; Kaihura and Stocking 2003; Gyasi *et al.* 2005) and two other closely related books (Brookfield 2001; Stocking and Murnaghan 2003). Several of the farmer networks that our collaborators developed between 1998 and 2002 were still active in 2005. Some of the 'experts' also had things to say to governments. In Brazil, one became an adviser in rural development and conservation to the governor of the state of Amapá, and declared that: '*para conservar no precisa poner cerca, precisa mudar a mentalidade da gente principalmente dos politicos*' (To conserve, you don't need to build a fence; you need to change people's thinking, principally that of the politicians) (Pinedo-Vasquez *et al.* 2003: 43). But the *politicos* have to satisfy a range of constituents to many of whom conservation is not the first priority.

Together with climatic change (and sometimes in competition for political interest and funding), the biodiversity 'crisis' has become a global issue in a way not paralleled by other environmental concerns (Zimmerer 2006). The major, and still ongoing, expansion of areas protected in one way or another, mainly in the developing countries, is largely driven by a global elite and supported by a large literate and television-viewing public, especially in the richer countries which provide most funding. Within those richer countries there has been less pressure to expand conservation territories but a growing concern to see land, its cover and water managed as public goods. This has generated new constraints on farmers, but also provided new means of funding them. The object is a less damaged countryside, to be achieved through more environmentally friendly agricultural practices. The protection and enhancement of biodiversity occupy positions of rising importance in these movements. Unsurprisingly, they have advanced vigorously in densely populated Europe.

Safeguarding diversity and farmers' welfare in Europe

The first steps in safeguarding diversity were taken at different times and in different ways in each country during the period when large subsidies for cereal production were still enticing farmers to convert yet more long-enduring grassland into arable fields. German efforts to regulate farming practices in nature conservation areas led to fierce political conflicts based on farmers' 'inalienable' rights. The principle of compensation for the loss of such rights was ultimately accepted, but a more positive approach was pioneered elsewhere. In Britain, the Wildlife and Countryside Act of 1981 required farmers intending conversion of land protected for its natural values under a 1949 Act, to notify such intention and if permission was refused the nature conservation authorities were required to compensate the farmers. This meant that they had to buy out the large potential subsidy thus forgone, an obligation they lacked the funds to meet. In 1984 a marsh-grazing area in the Norfolk broads which the owner-farmers intended to drain and plough provided a critical test. A hurriedly introduced scheme, funded by the UK Treasury, offered the farmers a flat annual payment to continue the low intensity pastoral use of the land, which they accepted. The incentive principle thus established became the model for most European agri-environmental schemes (AES) for the next 20 years (Latacz-Lohmann and Hodge 2003). While command-and-control did not vanish, farmers were thenceforth paid for conservationist management on 'sites of special scientific interest', and after 1986 in all nationally declared 'environmentally sensitive areas'. Meanwhile the German government had offered a nation-wide scheme supporting farmers either to reduce use of pesticides and fertilizers, or to convert to organic agriculture. Britain followed this model with the introduction of the Countryside Stewardship Scheme in 1991.

In the uplands rising costs were making farming increasingly unprofitable, especially for the smallholder farmers who predominated in these areas. Yet the high scenic quality of the uplands owed much of that quality to being farmed, and farmed in 'traditional' ways. The uplands occupy more than a third of the total agriculturally used land in Europe. The problem was rarely damage from high intensity farming. More widely it was the threat of partial or total abandonment and reversion to unmanaged wild. To Americans, who have seen much larger areas in their eastern states revert to woodland during the twentieth century, this might not seem a problem. To Europeans, who wanted to walk, cycle or drive in their managed countryside, it was. Biodiversity also entered the equation, since the managed upland farmlands provided habitat to a considerable range of valued wildlife. Of particular concern were the species-rich unimproved grasslands and traditionally mown hay meadows which were in rapid decline right across the continent (Macdonald *et al.* 2000). In France, local and regional initiatives were first supported in the late 1980s, especially in upland areas where landscape and biodiversity were seriously threatened by abandonment of grazing and scrub encroachment; the aim was to sustain productive use of land by offering support to the farmers (Simpson 2004). It became general EU policy to offer structural support to 'less favoured areas'. In upland areas of Britain, support has for years been offered specifically to hill sheep farmers (Gray 1997, 1998). In the uplands of the Czech and Slovak republics, where protected areas were first determined in the early 1980s, the initial objective of protection from modern intensive practices shifted abruptly in the 1990s to keeping the land in productive use, just as in western Europe (Prazan *et al.* 2003). Often the support offered is too thinly spread to convert unprofitable farming into something more sustainable.

The result of support is not always what environmentalists might wish to see. O'Rourke (2006) describes such a case in the Massif Central of France which in the nineteenth century was a classic area of cereal-and-sheep, or tree-crop semi-subsistence farming. The whole upland region began to lose population very rapidly so that by the 1950s national policy favoured its total afforestation. In 1962 the national government set about modernizing French agriculture in ways that included support for the less favoured regions by means of infrastructure investment and subsidy. Two areas of the Massif responded in different ways. On the limestone plateau of the Causse Méjan, farmers mechanized their pasture management, and upgraded their sheep flocks to produce larger milk supplies that were sold to nearby Roquefort cheese factories. To achieve this, they imported feedstuffs, advanced the lambing season to obtain winter production and confined the sheep for the larger part of the year. In the gorges of the eastern Massif, abandoned farm buildings were sold to newcomers seeking a rural way of life, who raised goats for cheese production on the farms. They also imported the valued hay of the La Crau region in the nearby lowlands. The whole region, with its farmers, later became a national park, but while some of the farmers, old and new, successfully added tourism to their activities, the more successful tourist operators were professionals. The sheep and goat farmers prospered with good prices but overwhelmingly as *hors sol* (off the land) entrepreneurs, aided both by regional development funding and supported prices. The question of who is to sustain the valued landscapes and their biodiversity remains unresolved. The farmers sell their produce as coming from a 'high nature value' farming region, but the high nature value is now more in the marketing than in production.

Measures to support diversification of the economy of the marginal 'less favoured areas' (LFA) were part of national policy before they became EU policy in the core countries of western Europe, including Italy (Chapter 6). Elsewhere, the linkages had to be built. In the Czech Republic collectivization of land in the 1950s extended far into such marginal regions as the White Carpathians of eastern Moravia where an old system of management of pastures and meadows, which had built up a rich biodiversity over centuries, suffered greatly as a result (Prazan *et al.* 2003). In 1980 a large part of the region became a protected area controlled by the Environment Ministry, which imposed quite rigid conditions. Under privatization in the 1990s only a few of the re-established landowners resumed their farms, and most were content to allow tenants to operate blocks sometimes as large as the former collective farms. The heirs of many of the pre-1950 landowners made no effort to reclaim title to marginal land at all, so that almost half the area is managed by a small number of farmers with very uncertain title. Under severe agricultural depression in the 1990s, livestock numbers fell precipitately and significant areas were abandoned and reverted to scrub. The Czech government introduced financial compensation for income losses due to environmental controls in 1997, and this has extended to support for the continued use of marginal land. There is no funding for the recovery of abandoned land. Very widely, the LFA support has been regarded as a social payment, and farmers find it hard to understand why, or for whom, they are urged to produce environmental goods such as biodiversity. In such a situation, where there is large emigration of young people, the protection authority and an NGO have had to begin working with small farmers who have tended to receive less support than the larger farmers. There was a perceived need for the type of multi-functional rural development support which, hopefully, should now become available through changes discussed below (Prazan *et al.* 2003).

Agri-environmental schemes (AES)

The conflict between productionist and conservation goals, neatly exhibited in the Massif Central, has been present throughout the evolution of the second, environ-mental, 'pillar' of the EU CAP, briefly introduced in Chapter 6. The conflict is sometimes between policy and reality, even within the policies themselves. Most countries began with regulatory approaches, with specific financial support confined to a very few areas. There were just four such areas in France in the 1960s, and a scatter across northwestern Europe by the 1980s. Only in 1987 did the CAP agree to reimburse 25 per cent of agri-environmental spending to those countries, almost all in northern Europe, that had developed AES schemes. Then in 1992 as part of the larger reform of the CAP, AES became compulsory for all member countries, co-financed from basic CAP funds (Bureau 2003). But AES funding remained only a fraction of total support, even as the demands were substantially increased by the Habitats Directive of 1992 and the evolution of a network of managed protected areas, the Natura 2000 network, that is expanding to occupy 15–20 per cent of the total area of the enlarged EU. These are areas of high natural value, especially from the standpoint of habitat diversity and wildlife. Management is modelled on the biosphere reserves of UNESCO. Only very small parts are totally protected; otherwise they are to be managed in a conservationist manner, involving require-ments much stricter than those of the 'codes of good agricultural practice' that have meanwhile become mandatory for all farmers receiving subsidies of any kind.

Codes of good agricultural practice were initially developed in response to the Nitrate Directive (Chapter 9). They were intended to be, and often are, little more than what good farmers would do anyway. Under the CAP reforms of 1999 and 2003, they became a 'cross-compliance' requirement. Inevitably, they had to be elaborated. As set out on paper in the form of regulations, for example in the five administrations of the British Isles, they read as formidably detailed documents subject to annual change,[2] specifying not only what is permissible, but also when. In some cases, they prescribe maximum input quantities and imply the possession of specialized equipment in order to achieve the standards. Allied to them are the requirements for managing set-aside land (Chapter 6), in which maintenance of habitat for wildlife and protection from pollution have become objectives equally important with the original control of production. Agri-environmental schemes, which can command substantial additional payments, should demand higher level management of at least the contracted areas within farms, but this is not necessarily the case where linear objectives such as hedges or walls, or the maintenance of local livestock breeds, are involved. Where there are very basic 'wide but shallow' AES, as in Austria, the conditions differ hardly at all from those that are now mandatory for all EU farmers.

The EU decision confirming AES as part of the CAP placed all decision making in the hands of member governments, and in federal and quasi-federal countries (Germany and the UK) each governmental unit has its own scheme. In some countries such as Spain and France, a province or département has control. Each unit is also responsible for determining the good farming practices that should lead to land being in 'good agricultural and environmental condition'. Even basic objectives differ. Some national schemes are specifically described as being for farm income support, while in others specific environmental goals are to be achieved. The basis of payment grew out of the early purpose to discourage intensification in the 1980s. This led to the continuing basis of production income forgone. In few instances has payment been determined by the full opportunity costs of the resource being managed, and thus far in none by determination of a 'fair return to the business of all the resources committed to the environmental scheme' (Centre for Rural Economics Research and CJC Consulting 2002: vi). It is generally agreed that AES should mainly involve purchase by the state and the EU of public benefits that the market does not buy, but there is semblance to simple transfer of an ordinary subsidy payment into the WTO green box, and in some countries also into funding of regional development objectives.

The period since 1992 has been one of depressed economic conditions for farms in all of Europe, but most especially in the east. The new AES opportunities were taken up with considerable alacrity and in some cases applications exceeded the number of contracts that could be awarded, but the extent to which this represents a general conversion to environmental goals remains very uncertain. Results are certainly good in some places, but much less so in others. Most evaluations have been short term and subjective, paying limited regard to environmental outcomes. One review of the available (mostly less than rigorous) evaluations of biodiversity effects showed, on balance, rather unsatisfactory results (Kleijn and Sutherland 2003), but it was too early for a firm conclusion. More valid was the advice that evaluation programmes should be an integral part of all AES.

The European Agricultural Fund for Rural Development: opportunities and conflicts

The 'second pillar' of the CAP has evolved at a great rate, both within countries and internationally. Although in effect initiated in 1992, it was formally established only in 1999. In 2003, when the definitive decision was taken to move basic support from commodities to a single farm payment uncoupled from production, it was also decided to rationalize and widen the 'second pillar'. As decided in 2005, the 1999 'rural development regulation' would be replaced by a European Agricultural Fund for Rural Development (EAFRD) from 2007, in which an economic emphasis is significantly restored. The new EAFRD has three main 'axes': improving the competitiveness of agriculture and forestry; improving the environment and the countryside (including the management of land in Natura 2000 areas); and supporting diversification and the quality of life in rural areas. These arrangements are to endure until 2013 and, as a result of a fierce debate that went on throughout 2005, with no increase in the agricultural budget as a whole in this time.

These compromise arrangements, and their significantly different application in member countries, reflect the collision of strongly contrasted forces. On the one hand is a large group of interests, including agribusiness as well as farmers, and supported determinedly by France and some Mediterranean countries, which seek to retain as much as possible of the market and income support that has so strongly favoured large-scale arable and livestock production. On the other hand is a rather uneasy coalition of interests seeking reduction – even elimination – of the CAP in its present form, with public support to farmers confined to payment for non-market environmental services, most strongly pressed by the British government (UK Treasury and Defra 2005). Somewhat loosely allied with the latter group is an international set of environmental and small farmers' movements, variously seeking a more socially fair distribution of funding and far greater emphasis on environmental issues than the funding that is likely to be available under the compromise will permit.

In southern Europe especially, rural development after 1999 was seen in terms of continuing modernization, and there was limited interest in AES, except for conversion to organic production. Plans for the EAFRD period continued on the same lines, with an emphasis on infrastructure, including even motorways in sparsely peopled regions such as the western Spanish province of Extremadura.

Problems like the disintegration of the ancient *dehesa* pastoral system in this province, following the end of transhumance along the old livestock migration routes, attracted limited concern except among ecologists and those worried by rural depopulation (Beaufoy *et al.* 2005). European directives were often seen as northern intrusions of little relevance in the Mediterranean south, to be complied with as necessary to keep up the flow of funding, but otherwise to be avoided by the classic methods of foot-dragging and non-cooperation. In an intensive irrigation area on the south coast, the EU offered market opportunities and protection, but its environmental directives were seen as at best a nonsense, and probably a serious threat to livelihood and prosperity (Izcara-Palacios 1998).

A state of confusion

With unresolved objectives go unresolved questions of means and methods. AES and rural development policies have undergone little change. The AES options are numerous and information for farmers is often scant; the implications of withdrawing the 20 per cent above-cost incentive payments in AES from 2007 had not been fully grasped in 2006 (Reichert 2006). Legal requirements in some countries are the objectives in others. Ireland, overwhelmingly a livestock farming country, made voluntary participation in an AES a condition of basic single farm payment, but in the absence of more precisely formulated general objectives of environmental management this condition is not easily replicable. It is very evident that the budget available for management of the Natura 2000 network is inadequate, even while the network is still being enlarged in eastern Europe. It is one thing to say that public support to farmers should be confined to payment for environmental services while the total agriculture policy budget should at the same time be reduced and another thing to prevent governments and regional authorities from finding ways to continue income support where they see it as needed and politically productive. As of 2007, the confusion continues.

Farmers and the environment in other lands

In the USA there has never been any doubt that environmental benefits should be purchased by the state, principally through the conservation reserve program (Chapter 9). Introduced to reduce overproduction and soil erosion, the reserved areas have been valued increasingly for their benefits in restoring patches of wildlife habitat (Allen and Vandever 2003). Beyond them, drainage of wetland has resulted in severe losses of wildlife habitat, as well as of pollution extending downstream into the seas. The 'swampbuster' provision in the 1985 US farm bill acted as an effective deterrent to unauthorized further drainage, and in 1990 a wetlands reserve programme was created with the effect that by 2000 more wetland had been re-created or newly created than had been cleared in the preceding decade (Claasen *et al.* 2001). This was followed in 1996 by the Wildlife Habitat Incentives Program, offering cost-share for re-creation of both wetland and upland habitat. States and well-funded private groups subsidized these resources, and in regard to a wetland in southern Illinois, Adams *et al.* (2005) analysed in some detail the conflicting interests involved. Finally, the conservation security program enacted within the 2002 farm bill introduced a three-tiered voluntary scheme on European lines for the first time. Biodiversity enhancement is only one of its three objectives, and soil protection still holds the most prominent place.

Among the most successful anywhere in the North Atlantic countries in terms of uptake by farmers is the flexibly designed Environmental Farm Plan Program of the Canadian province of Ontario, even with a level of payment well below the EU average for AES (Smithers and Furman 2003). It begins with a whole-farm assessment of environmental problems, initiated by the farmer, leading to actions to improve soil, water and nutrient management, wildlife species and their habitat. Funding, at 25 per cent of costs, is supplemented by free advice and by prizes for the best outcomes. By 2002, over half Ontario's farmers were participants.

Australia tells a different story, one of legal regulation with a very slow and reluctant move toward any sort of formal AES, but quite ready willingness to support voluntary efforts. Agri-environmental issues entered public debate in the early 1980s, when the neo-liberal policies of a new left-wing government were wiping out direct and indirect supports to agriculture in a manner second only to New Zealand (Chapter 8). A politician's initiative in the state of Victoria in 1986 created the first 'landcare groups' among farmers who would cooperatively work out solutions to their own perceived environmental problems (mainly soil erosion, control of feral animal pests and dryland salinization) with help from the state government. In 1989 the national government decided to sponsor a much wider network of landcare groups, and they grew rapidly to number over 4,000 by the mid-1990s, involving some 40 per cent of arable and livestock farmers (Cullen *et al.* 2003). Participation, even within the one state of Victoria, ranges from as few as 20 per cent to more than 80 per cent of farmers in an area (Sobels *et al.* 2001).

From the mid-1990s, long-standing concern over the extent of clearance of native woody vegetation for livestock and arable farming became a major political issue with pressure from conservation NGOs, and led to the strengthening of regulation preventing further clearance. Historically, land clearance was a requirement of leases in Australia and over more than a century from the 1850s very large areas were cleared or partly cleared. Already by the 1890s this was seen to be leading to rising water tables and salinization in some parts (Cunningham 2005). It continued and became more intensive as large-scale machinery and irrigation were employed, requiring removal of remnant trees. Two new public issues were added to that of salinization in the 1990s: biodiversity conservation and the realization that land clearance was a major contributor to Australia's large share in the production of greenhouse gases. In this matter, in which state governments have full rights, there has been solid reliance on command-and-control, inflexible in regard to farm plans and to the very rapid regrowth of some species on favourable land. Monitoring relies on satellite imagery (without systematic ground truthing) and on information from neighbours. Urban-based environmental NGOs have been very active in pressing for action against transgressors. An adversarial relationship has developed among many of the farmers affected, and a lot of land has been cleared illegally. A review committee, itself a bit overwhelmed by the ecological dynamism involved, found that the system had many defects and, while agreeing that landowners got useful benefits from the native vegetation on their land, felt the public goods of biodiversity, wildlife protection and reducing greenhouse gas emission were proper areas for public purchase (Productivity Commission 2004). Political promises notwithstanding, three years later the system was little changed.

By contrast with the rather dismal story of land clearance regulation, more support has been put into the landcare movement, especially where groups have acted together at a catchment level (Sobels *et al.* 2001; Cullen *et al.* 2003). Indeed, so much government support has been chanelled through the landcare movement that there are questions regarding its independence (Wilson 2004). Landcare has attracted some powerful corporate sponsors, which also can be selective about what they fund. There is scope for individuals as well as for groups to obtain funding from the very broad Natural Heritage Trust, formed in 1997 and renewed in 2002 with a larger budget. Grants, usually on an approximately equal cost-share basis with the recipient, are made for a great variety of agri-environmental projects, large and small, some with more obvious public benefit than others (at least on the basis of the published description). For farmers, none of this is easy, as Box 10.1 demonstrates.

BOX 10.1 LAND HUSBANDRY IN SOUTHEASTERN AUSTRALIA

Making a living from farming is hard enough with rising costs and stagnant prices as well as the vagaries of climate, but 'managing the environment' ramps it up a few notches. The grassy Eucalypt woodlands in southeastern Australia have been classified as an endangered ecosystem and the majority now is on land used for farming. Mostly, information about the best ways to go about managing a farm in the former grassy woodland country is not adequate. Scientific knowledge is based on fairly recent studies which raise more questions than answers: How many trees should there be? Should fire be used as some species require fire to propagate? How much grazing can be tolerated and is it essential for plant survival and regeneration? And so on.

For the farmer the question is – Where to start? There is grant money available but the policy priorities change annually so applications have to be targeted carefully. And there are different avenues, and trusts, and tiers of government and organizations. Each will fund a certain type of project – some for waterways, others for fencing, weed removal or revegetation, plus there are conservation volunteer organizations that can be accessed to help with some aspects of the work. Most funding sources require some contribution of funds or labour by the farmer. Membership of the landcare group can be the best way of unravelling the funding application process, or there are some government-sponsored advisers who have to spread their time over multiple government organizations and across regions.

Some old-timers know about the land and how to care for it. Knowledge comes mainly through watching over the years, through the good and the bad years, and seeing the responses of the land with its vegetation and wildlife. Newer-comers watch also and learn, but they start seeing it from an already degraded level. Some land can heal with little interference, but some requires massive efforts with earthworks to overcome compaction and erosion and reintroduction of most of the original species. And for the farmer then comes juggling the productivity constraints. The best examples of grassy woodland are also the best land for agriculture, so it is more often the marginal country that is set aside. But in severe drought even the marginal country may have to be opened up to keep the stock alive. Without rain the revegetation projects stall, risking compliance with the grant regulations. Continual management is needed over the years. Weeds relish being released from grazing pressure, kangaroos wreck fences and along with rabbits nip off new trees. It's a long slow process for the soil life and birds to return, for the trees to grow and the understorey to re-establish. It's mostly small farmers and those with other income who have the commitment to the country or the wherewithal to undertake long-term environmental management. Many have found that productivity has increased as a result of their innovation and experimentation, but the majority of farmers are not convinced.

A concluding remark about the agri-environmental wave

Whether under centralized direction as in the American and European schemes, or under the democratic anarchy of Australia's voluntary approach, AES have achieved significant results. One important aspect in Europe (discussed in Chapter 13) has been the major expansion of organic farming under AES funding. Everywhere, however, there is conflict of objectives, and productivist ones still dominate. The farmer's first objective is, after all, to make a living. Environmental questions are now important to most governments. In few developed countries do the mainly urban voters still believe farmers can be trusted to manage environmental issues, produce only clean food and take the sort of care of their livestock that owners of only pet animals regard as appropriate. The agri-environmental wave began as a counter-movement to the excesses of productivism and neo-liberalism. It hit a bad patch in only a few years when the BSE outbreak in Europe struck deep at sympathy for farmers. In so far as we do now live in a 'post-productivist' world, it is one that is more imposed on farmers than one arising from a fundamental change in the priorities of a majority of those on the land.

Notes

1   From the early 1990s until early 2005 we both worked for the biodiversity conservation project 'People, Land Management and Environmental Change' (PLEC), mentioned in the Introduction. It ran in 12 countries, and at its peak in 2000–02 it engaged over 200 scientists in a range of disciplines, their students and several thousand farmers. In this chapter, we draw on PLEC experience more directly than in the rest of the book. The project was set up by the United Nations University (UNU) in Tokyo, and its first few years were mainly spent seeking funds. There was a few years of funding from the Global Environment Facility (GEF), and a short postscript period in which our own continuing work was funded by UNU. It was not easy to get funds in the 1990s because we were seeking to promote farmers' own skills in the maintenance of diversity, and had to overcome considerable scepticism. However, the times were propitious. The initiation of PLEC just preceded the 1991 Rio de Janeiro Earth Summit at which the Convention on Biological Diversity (CBD) was agreed, and in 1996 the third 'Conference of Parties' to the CBD proposed strong emphasis on the conservation of biodiversity in agricultural production systems. In the same year, our proposals were finally ready for submission to the GEF. We had caught the biodiversity wave at its peak, and the successor agrobiodiversity wave as it was still rising. Starting in West Africa and going eastward, the countries in which we worked were Guinée, Ghana, Uganda, Kenya, Tanzania, Thailand, China (Yunnan), Papua New Guinea, Peru, Brazil (Amazonia), Mexico and Jamaica. The scientific coordinators lived in the USA, Britain and Australia. The project was administered in Tokyo and managed in Canberra.

2   For example, the English cross-compliance handbook for 2006 is a document of 74 pages, and its Scottish equivalent, more modestly subtitled 'notes for guidance', occupies 92 pages. Both cover a range of legal requirements as well as 'good agricultural practice'. The English document concludes with the 'helpful' advice that farmers who have participated in an AES that takes land out of agricultural use for good environmental reasons have also taken such land out of the scope for receiving the single farm payment, and urges such farmers to seek advice on their options. The need for whole-farm planning, creeping in through the 2003 reform but still far from general, is strikingly underlined, as is the need for an advisory service.

11 Collisions over land in developing countries

Mexico and Brazil

During the years of evolving agri-environmental programmes in the North Atlantic countries, events of another kind were attracting attention in developing countries. They included conflicts over access to land in Latin America, southern Africa and the Philippines and what has been termed 'de-agrarianization', especially in Africa and large parts of tropical Asia. This chapter and the next describe these changes and their context mainly through discussion of particular areas. The present chapter takes up the collisions over land in the specific cases of Mexico and Brazil, with briefer reference to land reform issues elsewhere. We use the opportunity to touch on the wider issues of land reform in the twentieth century, within which the period 1910–70 has been appropriately described as the 'golden age' of land reform (Bernstein 2004).

From the 1940s to the 1960s, land reform was popular with most governments and international agencies. Absentee landlordism, seen as both socially pernicious and economically inefficient, was a particular target in the many countries where land had accumulated in the hands of a small minority. In some reforms there was also the political aim of converting a dissatisfied rural population into a satisfied one that would support the status quo. This was an underlying purpose of the Japanese land reform launched in 1946–52 during the American occupation, and it proved remarkably successful in providing an enduring power base for the Liberal Democratic Party. The Taiwan case has been particularly praised for major reduction in poverty and inequality, and the sustained production improvement that followed (e.g. Griffin *et al*. 2002). It did begin with one big advantage – the government took over about 20 per cent of the cultivated area from its former Japanese colonial occupiers in 1945.

There have been major land reforms in several countries of Latin America, and some still continue. Except in Cuba, where the post-revolutionary reconstruction of agriculture led mainly to state farms in place of the former plantations, the common element in a diverse set of measures has been redistribution of land held by the often-absentee owners of haciendas. In several countries, including Mexico, there have been reversals in direction, but unlike those in the communist countries (Chapter 7) they have taken place over a fairly long period of time. In the case of Mexico, which we discuss in more detail, it is necessary to go back to the events of 1910–17.

Land reform in Mexico

There have been many attempts to explain the complex Mexican revolution, which began in 1910, in simple terms, with its shifting alliances and enmities, but probably more correct is that it was a mosaic of local rebellions. Each had its own objectives and they had greatly varied results (Liendo Vera 1997), but land reform was an element in all. In Mexican literature it is common to date the land reform from a decree issued by the 'first chief' of the revolution, Carranza, in January 1915 when he controlled only a small part of the country. This decree survived with little essential change to become Article 27 of the new constitution finally agreed among the victors in 1917. It authorized return of confiscated lands to the Indian and mestizo communities, but it also emphasized the protection of individual land tenure, except of the large haciendas. Scope for differences of interpretation was thus created at the outset. Most leaders of the victorious factions were owners (hacendados) of small or medium-sized haciendas; the drive for implementation of land reform came mainly from those who had emerged from lower economic horizons, outstandingly Zapata, whose uprising in the sugar growing state of Morelos south of Mexico City was the earliest and most enduring outbreak. This was almost entirely about regaining village land lost to the haciendas during the preceding decades.

In the first 15 years, reform moved forward or not according to the aims of each new president. Landownership through collectives, or *ejidos*, was formally legalized in 1922, but over the years between 1917 and 1933 only 7.6 million ha had been redistributed to them. Most of the land was still held privately, often under new ownership; private transfer outside the reform system, involving redistribution from hacendados to smaller-scale owners, was allowed under the 1917 constitution. Up to the mid-1930s, the reform had made little impression on rural inequalities. There was then a major surge in land reform under the socialist-leaning president Lázaro Cárdenas (1934–40), during whose term the total distributed was 17.9 million ha, benefiting three-quarters of a million campesinos.[1] The average size of holding made available to *ejidatarios* was much larger under the Cárdenas presidency than either before 1934 or since 1940. In the state of Mexico it exceeded 11 ha per *ejidatario*. From 1940 onward there was a steady and sustained reduction in the size of allocated holding (Liendo Vera 1997).[2]

Nevertheless, the reform continued. By the 1990s the total area in *ejidos* was around 40 per cent of the land area, and over 60 per cent of the rural population lived on *ejidos*, though many were landless. At no stage in the reform was anything much done about the many private holders of small farms and parcels of land. Among these latter, there were both subdivision by inheritance and transfer of parcels, and concentration by purchase into larger units. In the central highland state of Mexico, for example, the area held in private farms smaller than 5 ha declined by 41 per cent between 1960 and 1970 (Liendo Vera 1997). Stavenhagen (1970) showed that this was by no means only local: during the 1960s, when the number of *ejidatarios* increased by only 9 per cent, the number of private farmers decreased by 5 per cent due to renewed concentration of landownership.

Looking back over the land reform period in the state of Mexico, which at the end of the nineteenth century had 328 haciendas larger than 1,000 ha and 15 larger than 10,000 ha, 55.7 per cent of the land had been distributed to *ejidos* by 1958 (Liendo Vera 1997). Some of this, and a substantial part of what remained, had been transferred to campesinos by private transaction, and thus remained private property.[3] By 1988 there were 1,112 *ejidos* in the state occupying 76 per cent of the arable area (Liendo Vera 1997). The average allocation over the whole period of the reform was only about 2.2 ha per *ejidatario*, but there were also many private holdings of no greater size, and some much smaller.

The ejidos

The *ejidos* created by the land reform held blocks of land allocated to either existing or newly created communities. Parcels of land within them were allocated to individual members of the *ejido* group, the *ejidatarios*. Usually, there was also common land not allocated to individuals. The newly endowed *ejidatarios* had to prove a residential or other right, on the basis of which they received holdings of from a fraction of a hectare to as many as 14, the smallest holdings being in the most densely populated areas. These parcels were a usufruct grant from the state through the community, and carried the condition that they could not be sold, mortgaged or leased, though they could be inherited, and divided on inheritance. In practice, they could also be subdivided among family members. The land had to be worked in order to continue to be held, and for a time *ejidatarios* who migrated to the cities or abroad lost their land, and it was redistributed to others. This did not continue.

In law the *ejido* plots had to be worked only with family labour, but neither this rule nor the prohibition on renting remained effective for any significant period in any part of the country. Apart from those tenant sons of *ejidatarios* who would inherit the land on which they worked, there were numerous other related and unrelated persons in most villages. Most of them were landless labourers (*peones libres*) free to come and go, unlike their indebted predecessors in the haciendas, but still with very few rights. Rapid demographic growth led to major increase in the numbers of this new disadvantaged class. Reform allocation embraced some of them during the Cárdenas presidency, but otherwise little has been done for them. A few hundred of the many thousand *ejidos* that came into existence by the later 1930s were organized on fully collective principles, but even on most of these the individual *ejidatario* parcels came to be worked individually by the mid-1940s (Wilkie 1971). Over time, an increasing proportion of the field and associated work done on the larger *ejidos* came to be performed by the labourers. Most of the small early *ejidos* became communities of subsistence farms, growing the characteristic Mexican mixture of a range of maize varieties, beans and squash.

The state and the campesino in the later twentieth century

Mexico was governed by the same political party from 1929 until 2000. But the Party of the Institutionalized Revolution (PRI), like its less organized predecessors, was at all times a shifting coalition continually able to take in new ideas. Beginning with no real ideological direction other than a common sense of nationalism, post-revolutionary Mexico took on a socialist orientation in the 1930s, then absorbed prevailing ideas about state-guided development in the 1940s and 1950s, shifted from one side to the other of the political centre in the 1960s and 1970s and then, after a serious debt crisis and under prompting from international finance, absorbed the neo-liberal doctrines of the free-market economy in the 1980s. The change of presidency in 2000 to the business-oriented National Action Party (PAN) represented a further shift in the same direction.

The Mexican campesinos with small and middle-sized farms have been batted around through these national shifts, because the rural sector has at all times been heavily dependent on the government. Even when the land reform was mostly complete in the 1960s, between 2 and 3 million of the agricultural population had too little or no land, or else land of inadequate quality. Migration remained the only hope of betterment (González Casanova 1970). Although agricultural production increased rapidly from 1930 to 1960, most campesinos still lacked the resources to undertake fully commercial forms of production.

For a brief space in the early 1960s the country again became self-sufficient in the production of its principal food crop, maize, as it had historically been. This was achieved notwithstanding rapid growth in total population from 13.6 million in 1900 to 35 million in 1960. By 1970 it was 48 million and in 2000 the Mexican population reached 100 million. From 1960 agricultural progress began to lag, especially on the small farms. The gap in productivity between farms with 50 ha or more and the *ejidos*, which had closed significantly after 1930, again opened up. Post-Cárdenas governments were principally interested in the industrial development of the country, and while they did not neglect agriculture they were selective in their investments. In the 1940s and 1950s, there was heavy emphasis on large-scale irrigation especially in the drier north. In the late 1960s, Barkin and King (1970) wrote that rural development in Mexico was more prominent in rhetoric than in practice, supported with just sufficient action to assuage dissent.

To improve marketing, and keep consumer costs down, a national food marketing authority (CONASUPO, *Comisión Nacional de Subsistencias Populares*) was set up in 1965. CONASUPO was responsible for price supports, marketing and storage, imports, and distribution of basic foodstuffs. Until the 1990s it was involved in a large part of all business concerning maize, beans, barley, copra, cotton, rice, sesame, sorghum, soya, sunflower and wheat. Processing of grains and oil crops was an important activity, and a network of shops and storage facilities was maintained to serve the rural population.

These improvements came together with closer direction of the activities of the campesino. In the same period, emphasis was placed on improving credit. The specialized *Ejido* Bank was combined with two other national agricultural banks, thereby offering greater service to the larger landholders. By this stage, the private small farm sector had evolved to be the near equal of the *ejido* sector in terms of farm area, though fewer in terms of number of campesinos. During 1972–73, 86 per cent of irrigated land held by *ejidatarios* was in parcels under 10 ha; 84 per cent of privately held parcels were above 10 ha (Restrepo Fernández 1976). In the era of mechanization, the small farms were at a growing disadvantage. A main purpose of collectivization in the 1930s, carried out principally in irrigated areas, had been to improve water regulation and make possible the use of machinery, especially tractors and harvesters. Even though holdings in the irrigated, cotton growing Laguna district were no longer worked collectively after the early 1940s, the *Ejido* Bank still required campesinos to follow coordinated scheduling of their work as a condition of granting low interest loans, in order to make possible the collective use of water and hired machinery (Wilkie 1971). Under regimes of the later twentieth century, with greater emphasis on providing credit to commercial farmers, the same tight control was sustained although decentralized to the states. In the state of Mexico in the early 1990s credit for maize production depended on following the recommended regimes for fertilizer and herbicide application, or at least something close to them. Only two approved regimes existed, one for rainfed and the other for partially irrigated land (Woodgate 1997).

State support for the farming sector, which began in a small way in the 1920s, had become very substantial by the 1960s and 1970s. Prices were supported while high tariffs restricted imports; there was large infrastructure investment especially in irrigation, and direct subsidy of agrochemicals. Credit channels were widened and enlarged, and there was also a modest degree of technical support. While support already progressively shifted from the *ejido* sector to the private sector from the 1940s onward, the total volume of support continued to grow. The state food marketing company, CONASUPO, supported 66 per cent of the production (by value) of maize in 1981, its last full year of unfettered operation (Yunez Naude 2003).

The neo-liberal reforms and counter-reforms

After a serious financial crisis in 1982 it became national policy to bring about a shift to a market economy in the rural sector. After several lesser adjustments, big changes took place in the late 1980s. Guaranteed prices were abolished on all crops except maize and beans, and in 1993 these remaining support prices were abolished. The state fertilizer company was privatized, as eventually was crop-loss insurance. The purchasing activities of CONASUPO were sharply reduced, and its functions were either privatized or halted. In 1999 it was finally wound up (Yunez Naude 2003). Mexico adhered to an international Agreement on Agriculture in 1986, becoming a member of the WTO on its formation, and the North American Free Trade Agreement (NAFTA) was negotiated, the first such agreement between wealthy and poor countries that included agriculture. It came into force in 1994.

Although a fairly long transition period was negotiated for Mexican agricultural trade in NAFTA, in practice the permitted tariffs on imports above quota levels were not charged, and the transition period was truncated. The financial crisis of 1995 and the consequent devaluation of the Mexican peso accelerated this process, so that national prices became equal to the much lower prices of imported subsidized American grain by the end of the decade (Nadal 2000). All these changes, especially the fall in the price of maize, caused considerable distress, and there was a political reaction. Large demonstrations by both private and *ejido* farmers took place in Mexico City in late 2002, and an organization called *El Campo No Aguanta Mas* (the countryside can stand no more) was formed in early 2003 to lobby for policy revisions.

Intendedly, the most drastic reform, more properly a counter-reform, took place in 1992. The constitution was amended not only to terminate redistributive land reform, but also to remove the already often-flouted controls on *ejido* land. The amendment specifically permitted sale of the parcels without restriction. Most other provisions had long been anticipated in practice. Even land rights were transferred fairly freely within the community. Sale had taken place by informal arrangements, commonly described as 'gift' to anyone enquiring. Although the expectation of the reformers was that Mexican maize and wheat producers would be able to shift their activities into more profitable crops in which low Mexican labour costs would give the country a comparative trade advantage, the neo-liberal PRI governments of the 1988–2000 period did not leave campesinos entirely unprotected. In 1991, a new agricultural marketing authority (ASERCA, *Apoyo y Servicios a la Comercialización Agropecuaria*) was set up within the Ministry of Agriculture to intervene selectively in commodity prices. Then in 2000, ASERCA reintroduced direct market support, initially only in 'remote regions'. After the demonstrations of 2002, this support was applied in all parts of the country (Mayrand *et al.* 2003).

Since the start of NAFTA in 1994, ASERCA has administered a new programme of direct income supports (PROCAMPO, *Programa de Apoyos Directos al Campo*). PROCAMPO was set up as a transitional support for campesinos affected by price reductions and it is to end in 2008. Although a commitment was made to keep the level of payments constant in real terms, the financial resources made available had decreased by over 30 per cent by 2001 (Mayrand *et al.* 2003). PROCAMPO does not support traditional mixed cropping. A second fund supported changes in cropping but, except where contract farming has taken hold (Chapter 5), this has been rather less successful, and it seems that many campesinos have been using their new incomes from both these programmes to buy the fertilizer and other inputs they need for maize production.

Much has been written about the devastating effects of the modern agrarian reforms on Mexican campesinos, especially since American maize has been freely imported. However, the American supplies are mainly of yellow maize, not the preferred white and coloured varieties. While there has undoubtedly been a shift toward cultivation of yellow maize as seed is imported through commercial channels, there has been no major reduction in the area and production of maize in the country as a whole. Received prices for maize have fallen substantially, while the cost of commercially made tortillas has risen sharply, but there has been only a very limited shift into production of alternative crops. Widely in Mexico, the response to declining output prices and rising input costs has been an unexpected stability, and even increase, in the production of maize (Nadal 2000; Barkin 2002). Campesinos have been helped in their stubborn determination not to give in by a consumer revolt against factory-made tortillas, made increasingly of imported

BOX 11.1 AN UPLAND VILLAGE IN THE STATE OF MEXICO

The principal PLEC site in Mexico was San Pablo Tlalchichilpa, in the municipality of San Felipe del Progreso, about 100 km northwest of Mexico City, in the district of Toluca, the capital of Mexico state. It was part of the homeland of the Mazahua people before the conquest and the present people are either Mazahua or mestizo. In the early twentieth century the Toluca district had almost half its area held by large haciendas, and since the 1930s it has had a correspondingly high concentration of *ejidos* (Liendo Vera 1997). By 1991 San Pablo had just over 2,000 people (Castelán Ortega *et al.* 2003). The Toluca-based PLEC team also examined a valley area several hundred metres lower, in which easier terrain and better soils, better access, some limited irrigation and fewer problems with frost, permitted a more productive mechanized agriculture, yielding 3 to 4 tons per hectare, much more than the solely rainfed upland (Arriaga Jordán *et al.* 1997). The *punta de riego* irrigation offered one or two small irrigations a year, allowing campesinos to plant maize from one to two months ahead of the start of the rainy season.

The campesinos of San Pablo produced the bulk of their own food, and most farmers sold some produce, mainly maize, beans and vegetables. Small quantities of maize were sold mainly to middlemen, who provided cash on the spot. Otherwise they received cash mainly by working for larger farms or in Mexico City, especially in the construction industry. In the lower valley many young people had jobs within daily commuting distance (Arriaga Jordán *et al.* 1997).

Among the *ejido* plots there was individually owned land, rented land and loaned land. The largest farm studied, 7 ha, was made up of land tenured in four different ways and most sampled households at San Pablo had some privately owned land (Castelán Ortega *et al.* 2003). The farming system adopted by the newly endowed *ejidatarios* in the 1940s was the *milpa* system inherited from pre-conquest times, in which maize, beans and squash were cultivated together often with minor crops, and edible leafy weeds (*quelites*) were collected (Brush *et al.* 2003). Also grown were barley and wheat, faba beans, lentils and peas, and oats, alfalfa and vetch for livestock. Domesticated livestock were a major colonial period addition to the system. At San Pablo, but not in the lower valley where the hiring of tractors has become important, work depended heavily on the use of livestock (Arriaga Jordán *et al.* 1997).

The uplands around San Pablo were initially simply colonized by some of the *ejidatarios*, though until 1992 they were still regarded as common property. The area studied had been used for cropland before 1960 but only then was a settlement made and the home fields around it (*solares*) set up. A *solar* has much higher agrobiodiversity than the *milpa*, and chili, amaranth and a range of vegetables were cultivated among medicinal and ornamental plants, other crops and trees.

Since the 1970s, farmers concentrated increasingly on white maize, and the former *milpas* became monocrop maize fields using agrochemicals that were heavily subsidized until the later 1980s. The *quelites* were largely wiped out by herbicides. Often the *milpa* was reduced to being a part of the *solar*. Commercialization of maize was moderately successful in the lower valley, but much less so around San Pablo (Arriaga Jordán *et al*. 1997).

In less than 50 years the hillsides had become seriously degraded. Since the counter-reform in 1992, some individuals have made major efforts to restore their plots (Castelán Ortega *et al*. 2003). Badly eroded gullies were infilled, fine pumice was added to the soil, sedimentation pits, walls and cut-off drains created, and neglected boundary vegetation replanted. They raised maize yields on two farms from 0.5 and 1.9 tons per hectare to 5.4 and 4.3 tons (García Fajardo *et al*. 2004).

yellow maize, and of dough not prepared in the traditional way with lime and ash. Hand-made tortillas from white and coloured maize varieties have commanded premium prices in what has become more than just a niche market since 2000 (Mayrand *et al*. 2003; Barkin 2002).

Collectively, support under the programmes that have taken the place of the old subsidies was estimated to provide 22 per cent of all farm receipts in 2002 (Mayrand *et al*. 2003). Three-quarters of these subsidies came in the form of price supports made necessary by the large imports of subsidized produce from the USA under NAFTA. By 2002 these imports were perceived to be jeopardizing food security in Mexico (Yunez Naude 2003). In making proposals on international trade in agriculture to the WTO in 2001, the new PAN government complained of the distorting effect of developed country subsidies and remarked that 'the actual use of subsidies depends more on the financial capacity of a country than on its contractual commitments. . . . If this difference in the application of subsidies is allowed to prevail, then the equity in trade we are trying to achieve will disappear completely' (excerpt from Government of Mexico proposals cited in Mayrand *et al*. 2003: 42). Devoted adherence to one-sided neo-liberal policies has therefore created serious problems not only for Mexican campesinos but also for their government.

The resilience of the Mexican campesinos

The fate of the land legislation of 1992, which was supposed to end the *ejidos*, offers a clear indication of the limits to intervention. All individual holdings were to be surveyed and registered, so that once mapping was complete the *ejidatarios* could hold their land as private property, including a defined share of the remaining common resources. They could sell, rent and buy the land of others. They could decide to terminate the *ejido* constitution and live as independent farmers. A system of land registration and regularization was initiated through a new programme, PROCEDE (*Programa de Certificación de Derechos Ejidales y Titulación de Solares Urbanos*) which was to have completed its work by the end of 1994. It remained incomplete in 2003 (Nuijten 2003; Haenn 2006). Problems of private land among *ejido* land, such as that described in Box 11.1, have been among its major difficulties. There were disputes with owners of neighbouring private lands, and with state governments over state land (Haenn 2006), as well as unresolved disputes with private holders who had managed to retain land in the 1930s (Nuijten 2003). Most registered land is still without full title.

The expectation of government was that the *ejido* sector would be privatized quickly, and disappear as a major element in the rural economy. The reform aimed to encourage consolidation of land, and to facilitate acquisition by national and foreign commercial enterprises. The reality was different. There were fundamental problems. Land is not simply an economic resource; it is a social asset in communities where many have none. It is regarded both as individual and as family property. While the old and new laws permit only single inheritance of the whole parcel, land has been divided within living families and on inheritance. Supposedly inherited only by men, it has gone to both women and men. Many of those registered as heirs when the *ejido* studied by Nuijten (2003) in Michoacán, western Mexico, was formally set up in 1942, were not the ones who actually inherited the land. Buyers and sellers often just go to a notary to register a still-illegal sale before formal title has been issued. There is widespread determination to retain the right to dispose of land as the holder wishes, not according to the limited concept of property rights enshrined in the law (Haenn 2006). The land reform had created what has turned out to be a very resilient structure, even if it made very few of its beneficiaries wealthy. With price supports also restored in some measure, it seemed by 2007 that the neo-liberal vision of a new rural Mexico was in need of some changes. The *ejido*, supposedly killed off by the 1992 legislation, has remained a remarkably enduring product of the revolution that Mexicans still regard as their country's finest hour.

Back to Zapata in Chiapas?

The extent to which the ideology of the revolution has survived is strikingly demonstrated by recent events in the far southern state of Chiapas, part of the ancient homeland of the Maya people of southern Mexico and Guatemala. Beginning two years after the 1992 legislation had supposedly ended the Mexican land reform, a small but long-enduring 'Zapatista rebellion' has become much better known internationally than the rest of the modern Latin American story, and needs to be discussed in some depth if its remarkable visibility is to be understood.[4]

Little of the 1910–17 Mexican revolution had taken place in Chiapas, but the land reform reached it in the 1930s and, contrary to views commonly stated in the proliferating literature, work by van der Haar (2005) makes clear that by 1990 half of all land in Chiapas was held by *ejidos* or communal groups of indigenous Maya. The state did not enter as strongly into other aspects of rural management as it did in central Mexico, so that the new *ejidos* became effective local government organizations, and later formed area-wide unions. The situation was more fluid than elsewhere and from the 1970s other forces came into play. There was timber to be cut, a new oilfield, and the reformist 'liberation theology' of the Catholic church was strongly represented along with newly arrived protestant sects. The church groups took up some of the many outstanding Mayan claims for *ejido* land, which government, local or national, had been allowing to lie unsettled for years (van der Haar 2005; Stavenhagen 2001).

To add to problems, the region lay in the sights of the global biodiversity conservation movement (Chapter 10). In southern Chiapas, a large forested area formerly inhabited by the Lacandon people, increasingly colonized by landless campesinos from central Chiapas after 1950, was designated in the 1970s for landless migrants from western Mexico under the country's last land-reformist president. In 1978 a large area, already with settler communities, was declared a biosphere reserve under the international scheme sponsored by UNESCO. There were rising calls to halt the deforestation for timber, an important source of income for many of the Maya and other settler communities.

Partly on the initiative of the Catholic church, Chiapas also received middle-class activists from other parts of Mexico. In the early 1980s, one set up a *Frente Zapatista* which took up the cause of the Maya people who were still being exploited and sometimes deprived of their land by politically powerful landholders and developers. The movement organized resistance against their private vigilantes, and also began to initiate occupations of both private and state-owned land that were subject to unsettled Maya claim. The counter-reform of 1992, already discussed, deprived the Maya of hope of further government support for their many outstanding claims to land, and this prompted action.

A poorly armed rebellion was launched in January 1994, seizing the principal city of southern Chiapas and several smaller centres. As an armed rebellion it lasted only 12 days, although land invasions continued and widened in extent throughout 1994. The defeated 'Zapatista army of national liberation' retired into the canyons of the partly deforested Lacandon region, and during a prolonged ceasefire embarked on negotiations with the government. They were protected from the overwhelming superiority of the national army by a remarkable wave of both national and international sympathy for an ethnic minority fighting for autonomy, its land and livelihoods.

Instrumental in generating this support was very skilled use of the Internet and the media, mainly by the anonymous north Mexican leader using the *nom de guerre* of Subcommandante Marcos. From the very outset, the Zapatista rebellion took place in the glare of national and international publicity. International conferences have been held at unspecified places in what remains of the Lacandon forest, where the 'army' is said still to be based. More than once, the masked (and therefore notionally anonymous) leadership led demonstrations far into central Mexico, to Zapata's old base in Morelos, and even into the capital. At least until lately, the movement enjoyed a good deal of popularity among a Mexican population unhappy with the trends discussed earlier in this chapter. It has also received strong financial support from international groups.

Negotiation, running through the life of two national governments, has been extremely slow. It focused on the issue of indigenous autonomy, and never really either resolved or got beyond it. Although the ill-armed Zapatista 'army' never again took the field, private vigilante groups did, and there has been persistent low-level violence. After an uncontested advance by the army in 1995 and the murderous action of a private vigilante force in central Chiapas in 1998, the Zapatista movement switched its tactics to the seizure of local administration, of which a parallel structure has, in part, been established. Most of the land occupations, covering more than 150,000 ha, have ultimately been 'regularized', mostly through the internationally favoured market-friendly method under which the land is bought by the state from the original holders, and then allocated to new claimants at an unstated cost (Bobrow-Strain 2004). The land reform objective seems to have been achieved, but prosperity has not followed. Meanwhile, government efforts to bring 'development' to Chiapas have also enjoyed little success, with a very poor record in attracting foreign investment. Chiapas remains the poorest state in Mexico, with a high rate of emigration to northern Mexico and the USA (Villafuerte Solis 2005).

In 2001, the newly elected PAN government took up unresolved issues posed by the presence of numerous Mayan settlements in the Lacandon region, including the biosphere reserve. A group of Zapatista municipalities took up the issue with characteristic eloquence (Ricardo Flores Mago'n 2002). The movement was on more treacherous ground with this, for many of its vitally important supporters in Europe and North America believe that reservation of land for conservation purposes is good.[5] The movement survives, and government has allowed it essentially to die on the vine. An 'obituary' for the uprising (Radu 2004) may be premature, and some supporters continue to laud its successes, especially in raising the status of the Indian minorities. But as a social revolution, the Zapatista movement has never fulfilled the dramatized hopes of many of its early supporters (Brass 2005). Perhaps it never set out to do so.

Brazil

Events in Brazil have been far larger in scope than those in Chiapas. Up to the mid-twentieth century the structure of the production systems that had endured from the nineteenth century (sugar, coffee and rubber) remained broadly intact. Except among European settlers in the far south, beyond the coffee regions, there was no peasantry as such, the Amerindians having remained subsistence-oriented shifting cultivators in the small areas that remained to them. As briefly outlined in Chapter 8, most land was held in large estates, on parts of which farm workers had tied rights to cultivate land on a basis very like the debt-peonage of pre-revolutionary Mexico (Martins 2002). In the mid-twentieth century, technical and social changes in combination led to abrogation of these rights. The semi-proletarian workers were converted into full proletarian workers, and in the south peasant land was resumed for mechanized wheat and soya cultivation. Social and political movements opposing these trends were swiftly extinguished after the military coup of 1964.

The urban employment market for unskilled workers dried up at the same time as land ceased to be available for displaced farm workers, leaving only the informal sector for the many job seekers. When the military regime embarked on a major programme of colonization in the Amazon region and in the west, this brought colonists there into conflict with indigenous Amerindians. Both groups lost land and many lives to large corporations, some of them multinational, which sought both land and compliant labour. The situation was further complicated by the fact that very large areas in the west and north of the country were held on dubious and in some cases fraudulent title, often with conflicting claims. Welch (2004) tells of a case in which 400 tenant families cleared forest, planted cotton and settled on a leased part of such a speculative land purchase. They were ordered to quit after the leaseholder found better things to do with his time and money. The tenants stayed put, but were attacked and their crops and houses destroyed by the military police in 1973, and again three more times before 1980. The land, which had never been properly titled, was reclaimed by the post-military government in 1986 and then finally allocated to its occupants.

The 1970s was the period of most severe repression in Brazil. Church groups (principally Catholic but also Lutheran) used their relative immunity to take up the issues on behalf of rural squatters, supporting them mainly on 'human rights' grounds, both politically and on the land. Even the dictatorship created a fund for rural credit, services and relief in 1971, allocating money through unions which the government thus sought to co-opt, a step which led to the creation of many more unions. This *Funrural* survives, whereas other direct credit systems for small farmers, set up after the end of military rule in 1985, have perished in the hands of neo-liberal policy makers advising democratic presidents. As the military government became weaker in the late 1970s, opportunities arose for a loose coalition of church activists, emergent leaders among the displaced rural workers and urban supporters to embark on a new and daring course of occupying uncultivated areas of large estates and government land. The *Movimento dos Trabalhadores Rurais Sem Terra*, the movement of landless rural workers, or MST, was born and has spread from its original area of operations in the south to the whole of Brazil.

MST has overcome a great deal of local, regional and national opposition, through organization, a non-violent approach and skilful diplomacy. Post-military governments, even that of former dependency-theorist F.H. Cardoso, sometimes employed repressive measures. One lethal intervention in Cardoso's time backfired and increased both popular support and MST membership. Although no longer closely allied to the church organizations, MST followed a consistent policy of refusing any linkage to a political party until the political rise of the workers' party under Luis Inacio 'Lula' da Silva, who became president in 2003. Well-planned land invasions, on a scale sufficient to resist or discourage the landowners' vigilantes, have when successful been followed by the setting up of settlements for the production of crops and livestock (Harnecker 2002). Many of its leaders, and even the present left-wing government, might have preferred a consistently collective approach to the management of these settlements (Navarro 2005). Instead, MST adapted to the wishes of its member settlers, and evolved a mixture of individual farms and cooperative enterprises ranging from production, through processing to marketing, with only a small number of collective farms.

MST has financed considerable mechanization, livestock breeding, processing enterprises, education, health and social services by internal taxation, variable access to state and national credit, private and international support, and the *Funrural*. By 2005 it had over half a million member families (over two million people), the majority of whom are now settled on the land (Harnecker 2002; Welch 2006). In the years since its formal foundation in 1984, MST has substantially raised the share of family farming in the Brazilian rural economy, and acquired every appearance of being an enduring political and social force. However, its modern alliance with the Lula government has not brought the large increase in government-sponsored land reform that was expected, and even promised. The annual rate of legal land allocation to settlers under Lula has been only half that achieved under Cardoso, and the regime has increasingly favoured the flourishing export-oriented larger landholders and agribusiness. In 2006 over 200,000 families remained camped along highways, and many drifted away to seek places in the revived urban economy (Petras 2006). MST is no longer the only body supporting the squatters, and the future remains unclear (Navarro 2005). MST, or at least the leadership, has always sought more general national land reform among its varied objectives, but this has never even approached achievement.

During the two decades of MST progress and substantial success, Brazil has become a predominantly urban nation, while its agriculture has undergone a notable renaissance around the export production of grain, soya and meat. This has happened under neo-liberal policies followed by both right and left parties once in power. Within this enlarging space there has been room for the emergence of a well-organized and disciplined population of family farmers, recruited almost wholly from landless agricultural workers. From the outset they have been ready to supplement their productive activities with off-farm employment in the expanding rural economy, as well as in the towns. As Martins (2002) demonstrates clearly, MST is a hybrid creation of squatters, peasants, workers and indigenous people critically supported by highly motivated members of the national – and international – bourgeoisie. Unlike the more diffuse and opportunistic Zapatista movement in Mexico, it has come together around a single project of claiming land in the rural economy for the rural poor, and for a significant minority it has succeeded. Without itself joining those who seek larger systemic changes in Brazilian society, MST has initiated new forms of relationship between the farm worker, the landholders and the state (Welch 2004; Navarro 2005). But as the economy continues to grow and the nation becomes more urban, the social demand for land reform has diminished greatly.

Modern land reform elsewhere: the Philippines and southern Africa

Internationally, the issue of redistributive land reform was supposedly killed off by the unholy alliance of neo-liberals and the neo-Marxists in the 1980s, but it lives on in academic debate and in the real world. International financial institutions have proposed market friendly reforms, on the basis of willing-seller/willing-buyer agreement. In the Philippines, where there has been a series of partial and ineffective reforms since the early twentieth century, a new scheme initiated in 1988 had, supposedly, redistributed two-fifths of the country's farmland to the rural poor by 2001. Much of this was very insecure since neither power nor control over credit have been redistributed (Gultiano *et al*. 2004). The most approved reform design has been 'voluntary land transfer'. Borras (2005), using information from the field and from sample audits, showed that a large share of voluntary land transfer has been reform evasion, especially through transfers to family members, and even to family members of the reform administration staff. There have been ingenious lease-to-own and leaseback schemes, under which the landlord can determine who will or will not benefit. Once the transaction is registered and added to the reform achievement, the landlord and beneficiaries together enter into a special leaseback arrangement, sometimes with multinational companies seeking to enlarge their pool of contract growers. The land price, and with it the lease rent, is often reduced below the real value of the land and is amortized over the leaseback period (Borras 2005).

There has been support for new land reform efforts from the International Fund for Agricultural Development, or at least from Michael Lipton writing for them (IFAD 2001). The World Bank has produced a new policy paper on land issues in which the desirability of more equitable land distribution is recognized, but the Bank's insistence on legally acceptable change and combination with other development measures are emphasized (Deininger 2003). We noted above the adaptation of this 'market-friendly' approach to realities in Chiapas, and in a different social environment in the Philippines.

The land reformist views, expressed by Griffin (1974) and Lipton (1977) more than a quarter-century ago, and termed neo-populism by their critics, were raised again in 2002 (Griffin *et al*. 2002). This time, they were put on the basis of greater efficiency, in the context of the contested inverse relationship between farm size and productivity. Provocatively published in a journal of Marxisant leanings, it yielded a reply in nine papers led by the editor (Byres 2004). The respondents found Griffin *et al*. (2002) to be utopian, reactionary and anachronistic. Griffin *et al*. (2004) responded.

Griffin *et al*. (2002, 2004) recognize the difficulty of land reform, and the rarity of its success. They are clear that confiscatory action is no longer internationally acceptable, and have no illusions about the very high cost of any true market-friendly reform. The charge of being reactionary depends on the reality of small farm efficiency, on which there is both positive and negative evidence. However, there is a real case for regarding neo-populism as anachronistic. It is very much associated with what happened in the twentieth century. Griffin *et al*. (2002) write that 'land reform is not a technocratic exercise; it is a transforming political event'.

It might once have been, as it was in Mexico between 1910 and 1940, but the critical issues now are different, and neither in Chiapas nor Brazil has any 'transforming political event' taken place. The emphasis in the new Griffin presentation is not so much on land per se as on labour, where inequality in landholding creates a means by which labour becomes more readily available to the capitalist large farmer. Griffin *et al.* (2002) argue that 'the purpose of land reform is to rupture the system of labour controls and bring to an end the monopoly and monopsony powers of large landowners'. Labour is not that easily freed in a situation where most workers and their families are indebted.

Bernstein (2004), who makes the most insightful contribution to the new debate, suggests that while the old 'agrarian question' of capital might well be passé, there could well be a new agrarian question arising from the apparent inability of the newly globalized economy to provide a living wage to all members of its highly fragmented labour force. This fragmented labour force now often works across urban and rural sites, non-agricultural and agricultural sectors, wage employment and self-employment, and often also across national boundaries and even continents. He suggests that modern land invasions reflect a new value given to land as basis for security by the failure of employment to provide for the needs of this 'footloose' workforce (Bernstein 2004).

Bernstein, following Sender and Johnson (2004), shows how 'market-friendly' reform in post-apartheid South Africa had reached only a small part of the rural poor, many of whom had meanwhile lost their jobs, and their homes on the farms, or had been forced into casual employment organized by contractors (Kritzinger *et al.* 2004). In Zimbabwe, a legalist approach yielded even less, by 1997 redistributing only around 3 million ha of the 15.5 million occupied by 6,000 large white-owned farms in 1980. An important part of this was land abandoned by its owners during the guerilla war of the 1980s (Bernstein 2004). Then a massive set of land invasions began in 2000, with the backing of government, reducing the still mainly white-owned farms to only 1 million ha in about two years. Called by the Zimbabwean government 'fast-track resettlement', this almost uncontrolled process wiped out the jobs of more than 300,000 farm workers, many of them immigrants or the children of immigrants. About half of these people have remained where they were, competing or collaborating with the land invaders for livelihood.

Conclusion

To Bernstein (2004), this new 'agrarian question' of labour does not support Griffin's case for a new wave of managed land reform, but presents itself as a moral and policy issue that needs to be faced in the context of the globalization taking place within capitalism. It also suggests that redistributive land reform within capitalism cannot yet be consigned to the 'dustbin of history'. The remarkable resilience of the Mexican campesinos should be placed in this context, as should the considerable success achieved by the Brazilian MST. The classic form of the 'agrarian question' becomes unstuck when those 'released' from the land can no longer readily be absorbed in other sectors of the economy.

Notes

1   As in Chapter 8, except where we are referring to the owners of large properties
    (hacendados), or specifically to the holders or users of *ejido* plots, who are *ejidatarios*,
    we use the Spanish term campesino (mentioned in the Introduction) to describe all
    Mexican farmers of small and medium-sized holdings. We also treat 'hacienda' as a
    readily understood term.
2   The literature on the Cárdenas reforms is strongly coloured by the contrasted political
    views of the writers, especially Mexican writers. While most writers see it as the peak
    of the reform period in Mexico, others (e.g. de Anda 1974) see it as a totalitarian
    deviation from a reform directed toward creation of a population of privately owned,
    and commercially oriented, farms. Writing while land reform was still a contentious
    issue, Cumberland (1968: 294–307) provided a balanced account.
3   Liendo Vera (1997: 399–400) goes on to remark, on the basis of a study of the records
    for the State of Mexico up to 1989, that 'La realidad fue tal, que la distribución de la
    tierra utilizando dos medidas: una, la grande, para el hacendado. La otra, la más
    pequeña, para el campesino' (The reality is that the redistribution of land took place
    using two measures: one, the larger, for the hacendados, the other, much smaller, for the
    campesinos).
4   The rebellion has a huge literature, including several books of which we cite only one
    (Harvey 1998). It was also the subject of a whole double issue of the *Journal of Peasant
    Studies* in 2005, from which we cite a few articles in the text. Perhaps the most cogent
    single account of events up to the change of national government in 2000 is by the
    veteran Mexican agrarian sociologist Rodolfo Stavenhagen. Published rather obscurely,
    it is more readily found on the Internet (Stavenhagen 2001). Surprisingly, it is little
    cited in the wider literature.
5   The history of this now largely deforested Lacandon region, close to the Guatemalan
    border, is itself a matter of contestation. The well-researched Zapatista document
    (Ricardo Flores Mago'n 2002) maintains that the Lacandon people became extinct in
    the late seventeenth century, and that the present titular inhabitants are in reality Carib
    migrants who had been co-opted by the state to hold a big area of land in the interests
    of commerce. On the other hand, it would seem that some of the Maya communities
    have been in the region since the 1950s. Eastern Chiapas as a whole contains numerous
    antiquities from the old Mayan empire.

12 Contrasted 'de-agrarianization'

Africa and Asia

Sub-Saharan Africa

In tropical sub-Saharan Africa, up to 90 per cent of all farm land is held under communal tenure, notionally under the authority of chiefs (Chimhowu and Woodhouse 2006). This does not mean that individuals lack strong rights to defined areas, nor does it mean that there are no major inequalities. Nonetheless, except in southern Africa the major modern events discussed in the rural literature have not been conflicts over land as in Latin America; they have been about the diversification of labour inputs and sources of income. Colonialism began this diversification, with its inescapable demands for tax, export commodities and labour. Diversification went much further in the late colonial and early independence period, and then was dramatically accelerated under the 'structural adjustment' policies imposed in the 1980s and 1990s.

In theory and rhetoric, the object of the pressure on governments to deregulate their economies, and to dismantle all institutions that might 'distort' the free market, was to create an improved market for African produce. What happened was different. Government management of export marketing did not prove easy to replace. Prices did not improve and even declined. Subsidies on inorganic fertilizers, widespread in the 1960s and 1970s, were an early casualty, adding to costs. The introduction of fees for health services and education, which were free before the 1980s, helped fuel continuous, year-round demands for cash. Agricultural incomes from the export crops, on the other hand, came only at intervals, and incomes from marketing food to the urban population, where available, were insufficient. It became necessary for farm families to intensify the search for means of survival and income accumulation.

The consensus of a varied and very unequal set of household-level inquiries in the 1980s and early 1990s had found reliance on non-farm incomes to be in the range 30–50 per cent of total income (Ellis 1998); the late 1990s showed a different picture. A multi-country survey by Ellis and Freeman (2004) now found a wide range around an average split close to 50 per cent. Using another set of household surveys in an overlapping group of countries, Jayne et al. (2005) found lower levels of reliance on non-farm income, but the broad patterns were much the same.

Looking more specifically for non-farming activity, in a six-country study, Bryceson (2002) found non-agricultural income to be providing 60–80 per cent of total household income at her sampled research sites. Particularly remarkable was the increased involvement of women in non-agricultural work, in or away from the home. 'Sales of prepared snacks, beer [except in Muslim regions], hair plaiting, petty retailing, prostitution, knitting, tailoring and soap making are a few of the many services they sell' (Bryceson 2002: 732). At one site in northern Nigeria, there were 89 different female-operated non-agricultural activities, compared with 79 for men, yet in no country had any significant proportion of women relinquished their traditional roles in household domestic labour and subsistence. Meantime, the declining value of the 'traditional' African export crops undermined the role of men, who were primarily responsible for them. The role of the household head as dispenser of funds had widely diminished as a result.

'De-agrarianization'

In first drawing attention to and analysing these changes, Deborah Bryceson (1996: 99) coined the term 'de-agrarianization' and defined it as 'a process of (i) economic activity reorientation (livelihood), (ii) occupational adjustment (work activity), and (iii) spatial realignment of human settlement (residence)'. Commenting, Ellis (1998: 10) focused on the income aspect in summarizing a hypothesis that 'rural [sub-Saharan Africa] is becoming steadily less agrarian as manifested by rising reliance on non-farm income in rural areas'. In other writings, Bryceson (1997, 2002) specifically described an African variant of de-agrarianization which she termed 'de-peasantization', meaning the end of the colonially created peasantry which combined cash production with subsistence.

A more nuanced picture emerges when village populations are divided into different levels of asset-holding households, following the livelihood-assets methodology developed in the 1990s and elaborated by Carney (1998) and Scoones (1998). Villagers with lowest incomes, least land and livestock, least education, poorest housing, smallest working households and fewest working tools are the ones most reliant on subsistence production. The off-farm work they are able to obtain is unskilled, whether working for larger farmers or in town. Those better endowed with assets are far less dependent on their own subsistence production, and are involved in more rewarding off-farm activities, in businesses or as salaried employees of government. In general, too, the poorest and the least poor are both more involved in off-farm work than the middle-income group of farm households. This widely encountered finding led Ellis (2005) to distinguish between off-farm employment due to pressures of survival, characteristic of the poorer farmers, and the seizing of opportunities characteristic of the relatively rich.

De-agrarianization is a powerful concept and it has been taken up quite widely, but it perhaps covers too much for one overriding interpretation. We find reason to question it. Such reorientation of livelihoods and incomes occurred much earlier in South Africa as a result of land policies designed to create enforced labour for the white farms and mines; modern changes have different cause.[1] There have been other trends as well. Self-provisioning has increased in importance as a survival strategy. One set of village surveys, in four countries, found that in 2001–02 from 57 to 97 per cent of food crop production was consumed by the households that produced it (Ellis and Freeman 2004). This is more like re-agrarianization than de-agrarianization. Nor is farmland losing its value. There has been acceleration in the evolution of 'informal' or 'vernacular' transaction in customary land, both for purchase and rent. Data are scant because the transactions take place outside formal land administration, and therefore escape national statistics (Chimhowu and Woodhouse 2006). Detailed local inquiries have, however, revealed a great deal of transaction in land (see also Chapter 3).

Close to a rapidly growing modern town on the maize-surplus southern highlands of Tanzania, the large village of Kinyanambo has been radically transformed, but by no means wholly away from farming. In a two-part paper focusing principally on the evolution of a land market, Daley (2005) shows how Kinyanambo was already a settlement of a few hundred people before it became the designated centre of region-wide concentration under the socialist villagization (*ujamaa*) policy of the 1970s. Kinyanambo continued to grow rapidly during the 1980s after socialism, especially because of its immediate proximity to the new town of Mafinga. Many of the new residents were principally interested in employment, business and trade, but a considerable number sought land for farming. Legally, land cannot be bought and sold in Tanzania, so they bought the 'unexhausted improvements' on land first allocated on customary title, and later paid village councillors 'sitting time' to formally register the transfer. The population reached 5,341 in 1999, and the largest group were the newcomers who had arrived during the 1990s.

By the end of the century, some of the customary holders who were richest in land were now among the poorest in cash income terms. Older women, often looking after grandchildren orphaned by HIV/AIDS, earned money from making and selling maize beer and the fermented juice of bamboo sap, and farmed mainly for self-provisioning. Though they might not admit it, their land was of value still. Land continued to be in increasing demand. Those with sufficient funds to buy it, and able to employ labour, could gain good incomes from the sale of produce in the large urban market in Mafinga, and through there to other parts of the country.[2] There are also good incomes to be made in skilled trades and in regular employment. Diversification of livelihood activities within households was due to the seizing of opportunities rather than from pressures to survive. Except for those no longer able to farm, almost everyone is now seeking land. Most of them expect to either buy or rent it from individuals. Land, and especially good land, has become very expensive even for those with good social and political connections. It can bring status and prestige, as well as incomes (Box 12.1).

In Nigeria, enterprising Yoruba farmers (Chapter 5) some distance from urban markets were diversifying as well as expanding their activities in response to rising demand for produce in Ibadan and other cities of southwestern Nigeria. In 1988 many of the younger generation of farmers had been away for work and training in the cities in a wide range of trades, but urban employment became much less rewarding once the Nigerian oil boom came to an end, and as structural adjustment

---

BOX 12.1 IMMIGRANT TENANT FARMERS IN SOUTHEASTERN GHANA

Scarcity of land has pushed up rents in many places, not least in southern Ghana where migrants from areas with poor access to the southern city markets have rented land for shares of one-third or even one-half of their main crops (Amanor and Diderutuah 2001; Gyasi *et al.* 2003). Over time, the more substantial of these tenant farmers began to acquire not only security but also respect among the people from whom they rented land. Through PLEC's very active group in Ghana, Brookfield had five opportunities to visit a cluster of Akan-speaking villages between 1993 and 2002. There were always some immigrant tenants and after 1996 they became the majority, with their own village settlements. Chiefs are of major social importance in this region, and formal meetings were always held in the chief's village until after 2000. Tenants attended, first as a small and silent group on the edges, then as active but quiet participants, and later as frequent speakers and contributors. The final formal visit (after the old chief had died) was held in a large tenant community. On the fringes of the meeting active trading in plant germplasm among the women made it hard for the characteristically long formal speeches to be heard.

---

had its characteristic impact on urban wages. Many returned to farming, often being helped by their fathers to re-establish themselves on the land (Guyer 1997). Whereas it had been expected that a very substantial de-agrarianization of the younger generation would take place, what has happened instead is an increasing special-ization in agriculture, as new commercial opportunities have arisen leading to on-farm diversification among the majority of farmers. New occupational groups have arisen, such as tractor owners and drivers, mechanics, vehicle operators, market traders and regulators. There is still insufficient division of labour for the skills learned in town to become the basis for viable entrepreneurship at home, so that the farm remains at the centre of the cash economy. Yoruba women have made many farms of their own since the 1970s, not for household use of food crops, but for money. But farming is only one of a repertoire of cash-earning activities that Yoruba women undertake (Guyer 1997).

Far from good market opportunities the situation is very different. In two areas of Burkina Faso, Brons (2005) found supplementary income from non-farm work to be important, but nowhere enough to lift households out of poverty. Brons makes the useful point that while farming is truly a household activity, non-farm work is individual, and the choices are individual. But opportunities are limited. In South Africa, Twyman *et al.* (2003) found that Tswana farmers in semi-arid northwestern South Africa and southern Botswana were certainly in trouble, but their basic problems were financial. Having been urged into cash cropping during two decades after 1980, they needed financial support to farm with tractors and fertilizers, and found it hard to carry over funds from one year to the next, so that they often had to sell livestock or work for wages in an increasingly constrained labour market. Households still wanted to farm, but some were not able to afford to do so. 'Smallholders in the Kalahari have had to diversify livelihood strategies in complex ways that combine pensions, welfare payments, off-farm income, remittances, loans, savings and so on' (Twynam *et al.* 2003: 16). This is not de-agrarianization, but simple hardship made worse by climatic and market variability.

'De-agrarianization' and poverty

In Africa, as in many other lands past and present, it is those farmers who have most successfully diversified their income streams who have best escaped the traps of rural poverty (Ellis and Mdoe 2003). Land productivity itself increases with rising total income, a finding replicated by 37 sample village surveys in Uganda, Tanzania, Kenya and even low-productivity Malawi (Ellis and Freeman 2004). The poor are employed by successful farmers who are often the main sources of employment in regions with few urban opportunities. Escape from this Lenin-like internal differentiation of the peasantry is not easy to achieve. There is a host of books, reports and articles advocating means to encourage increased rural production and marketing. There are many development projects with this as their central aim. Unfortunately, uptakes and outcomes have often been disappointing. Although cultivated area, production and livestock output have all increased in most parts of the continent, and although there is every possible reason to query the macro-level data, there is little reason to doubt the general conclusion of UNEP (2002) that per capita production has only locally increased and over wide areas has declined.

Escape from rural poverty does not lie solely in rural productivity improvement, important though this is, nor does it depend on a levelling of landholdings.[3] It requires a widening of non-farm and off-farm opportunities of a relatively scale-neutral type, particularly in marketing and in artisanal work where it is possible for the poor to accumulate small surpluses with which to enlarge their range of assets. Unfortunately, few of the national poverty reduction strategies address the institutional problems facing small farmers who seek to make such transitions, which are always easier for those households in which the head or another adult member has a higher level of education (Jayne *et al.* 2005). Not least among the institutional obstacles is the daunting problem of regressive taxation at the local level on small transactions. These are a feature of many African countries, even where these are not augmented by the exactions of local despots. They are reminiscent of the 'peasant burden' in modern China (Chapter 7). For Tanzania, Ellis and Mdoe (2003) set these charges out in some detail.

Pastoral people in conflict

Older competition for land among pastoralists and between farmers and pastoralists still persists in Africa. In western Botswana, the sinking of boreholes for water had the effect of privatizing the rangeland around these boreholes for the fortunate few who were able to claim the new water points (Rohde *et al.* 2006). In West Africa, the fierce droughts of the 1970s and 1980s led many pastoralists and farmers in the semi-arid regions to become what have been called 'environmental refugees'. In Senegal, some Fulani pastoralists had to move some 200 km south to the valley of the Ferlo where they were able to settle with caution on the fringes of the land of local pastoralists, taking advantage of the traditional and legal free access to water. They innovated by carting water from bores initially in the inner tubes of the large tyres used on construction machinery and later in more secure vessels, and were able to utilize distant pastures under lower grazing pressure. Disputes arose as new settlers found themselves being charged for legally free water and land. They were resolved by political lobbying and because the innovative new grazing strategies proved advantageous to the resident people as well as the settlers. Juul (2005: 123), who describes this successful adaptation, cites a comment by Agrawal (1998) on migration strategies in Indian Rajasthan, noting that success demands skill in finding 'one's way through a thicket of dynamic [land and water] property rights regimes'. In northern Mali, Toulmin (1992) described such a 'thicket' in which Bambara farmers had succeeded over a long period in confining Fulani and Maure pastoralists to a client status by sustaining control over water rights. Private wells, dug since 1960, attract Fulani herders who then corral their cattle on designated fields of the well owners to provide manure. A great deal of political pressure and negotiation even at national level has been needed to prevent Fulani from acquiring land and digging wells of their own, as the national law would permit.

Toward an African conclusion

Wherever there are sufficient data, it becomes evident that however bad urban conditions may be, poverty is worse in the rural areas (Ellis and Freeman 2004; Ellis 2005). In a paper of different intent from those cited above, Tiffen (2003) relates opportunity in African small-scale farming to levels of urbanization which, although rising, remain quite low by world standards. This is important because towns create a more reliable market for rural produce than has been provided by international exports. In a sense, the argument harks back to the regional development literature of the 1960s and 1970s, and perhaps especially to Johnson (1970). He argued that a network and hierarchy of towns was necessary for the full provision of services and markets in a developing country, and that many countries lacked a sufficient network. Tiffen saw sub-Saharan Africa as being in a stage of urbanization in which a substantial urban market capable of providing the main driver for rural surplus production is only now coming into existence on a sufficient scale. In a later paper she demonstrates the dramatic effect of urban-market expansion on livestock farming in a group of African countries (Tiffen 2006).

Some writers continue to argue that rural development based on the small farm is the foundation of economic development (e.g. Toulmin and Guèyé 2003, 2005). Ellis (2005) describes such advocates of technological improvement in farming, who seek to garner by new means what the Green Revolution failed to achieve in Africa, as 'agriculture optimists'. They see livelihood diversion as emerging from agricultural success. This remains national policy in some countries, most notably Ethiopia where the national development strategy is one of 'agricultural-development-led-industrialization'. But such rural-based strategies can do little more than make a contribution to poverty relief and are not conducive to transformative development.

In most of sub-Saharan Africa off-farm diversification first became substantial under colonialism, with its encouragement of labour recruitment for mines and plantations. The employment which provided income for farm families lay at a distance, and the agricultural surplus demanded was in the form of exportable cash crops. Except in parts of West Africa, as discussed in Chapter 5, it is only recently that provisioning of a large non-agricultural population has been required of African farmers – other than those called on to support armies in the subcontinent's numerous pre- and post-independence wars.

These wars cannot be disregarded, because many of them have led to very real de-agrarianization, as farms have been looted and burned, men killed, women raped and livestock butchered to feed the soldiers. The events in Rwanda and the eastern Congo since 1994 have received wide attention, and have a large literature. They seem likely to have been the most destructive of life and livelihood of any of Africa's numerous independence and post-colonial wars, with or without foreign intervention. But all the wars have been destructive. In many of them there have been strong economic as well as ethnic and political dimensions, especially in mineralized regions. The armies have lived off the land, have taxed the people heavily, and have sought both fighting and supporting personnel of both sexes and all ages from among the rural people. Added to these problems, and widely experienced in sub-Saharan Africa, have been dictatorial regimes that have demanded tribute and savagely oppressed any opposition.

It is not sufficient to interpret the last 25 years in Africa around any one set of linear processes. Nor can the future be seen as simply an extension of the past. Any rural development strategy that depends on exports to the developed countries has to confront a worsening of competition from which sub-Saharan Africa has suffered since the 1960s. Toulmin and Guèyé (2003, 2005) are two of many writers who call for removal of what they regard as unfair competition from better endowed countries which subsidize their own farmers. They argue also that development based on Africa's 'traditional' export crops (cocoa, coffee, groundnuts, cotton and palm oil) flies in the face of common sense given the weakness of their received prices. There would be less of a problem if the internal market, still growing rapidly through demographic increase, were better distributed. At present, a lot of the net buyers of food in Africa are the small farmers in rural areas: their activity diversification has the purpose of earning income to buy food which their own farms, constantly growing smaller through inheritance, are no longer able to supply. Following Tiffen (2003), Ellis (2005: 14) concludes that 'rural poverty reduction requires a much more rapid rural–urban transition than has been occurring in most [sub-Saharan African] countries over the past three decades'. In this view some of the indicators of de-agrarianization become positive rather than negative. Bahiigwa et al., reviewing the evidence, conclude that the best way to alleviate rural poverty in sub-Saharan Africa is to facilitate the more rapid growth of towns through infrastructure investment. 'Conditions in agriculture will automatically improve if rapid growth in food demand occurs in fast growing cities' (Bahiigwa *et al*. 2005: 120).

Southeast, south and east Asia

The term de-agrarianization has been applied also to Southeast Asia (Rigg 2001, 2003; Wilson and Rigg 2003). The generalized processes are not the same as in the original African presentation. Structural adjustment has certainly taken place in Southeast Asia under the same pressures from international financial institutions, but Southeast Asia has been less dependent on these institutions, and did not experience the same steep decline in the money product of its traditional agricultural exports, which instead have grown substantially. Nor did Southeast Asia experience any major set of financial crises such as took place in most of Africa and also Latin America in the 1970s and early 1980s, except for a short period in the late 1990s. This period apart, most of Southeast Asia has almost consistently enjoyed prosperity since the 1970s, based mainly on natural-resource exploitation and rapid industrialization. Yet a process of de-agrarianization, as defined above, has taken place faster and more thoroughly than in Africa.

As recently as the mid-1960s there was little sign that Indonesia, Thailand and Malaysia, the countries mainly discussed here, would soon embark on a path of rapid development. There were large and growing disparities of wealth, unrelieved rural poverty and growing urban poverty fuelled by a large immigration from the countryside. There was already substantial rural diversification, principally in trading and also in processing of some cash crop products for market. Trading systems were complex and very flexible, as Dewey (1962) and Geertz (1965) showed for eastern Java, and they operated with minimal need for start-up capital. A high proportion of traders were also farmers. In subsequent years, and especially in the 1970s, migration to urban areas became even more substantial, largely for employment in capital-intensive factories which, in Malaysia and Thailand, were already geared mainly to export. Integration of rural and urban economies went far beyond the supply of migrant labour. In 1960, the range of daily movement from most villages was limited to walking distance, or bicycle distance for a minority. Buses ran only on the main roads. The transport revolution that followed, mainly using small vans converted for passenger use, shared taxis and motor cycles, widened this range. When Brookfield first visited Yogyakarta in Java in the late 1960s, the surrounding roads were already full of cyclists and walkers each morning and afternoon; by the late-1980s they were crowded with motor vehicles of all kinds.

The same transformation occurred all over the region, and with it the later establishment of industries outside the cities proper. In Java and West Malaysia there had been significant industrial development in small towns as well as cities in the earlier decades of the twentieth century, a process led by migrants from southern China, but with significant national entrepreneurship as well in Java (Geertz 1963). The industrial expansion that took place after the mid-1960s increasingly involved the establishment of factories which drew large numbers of migrant workers from rural areas. Most of it was located in the cities, but the need for space demanded that factories also be established in locations outside the built-up areas. With rapid industrial expansion in the 1970s, new industrial estates were set up, not only in the immediate peri-urban areas, but well beyond them. These were mainly occupied by larger factories, often foreign-owned, leaving an acute lack of space for the more numerous small industrial establishments which had to find sites in or around the urban system (Chi and Taylor 1986).

Expansion and growing integration of urban and rural activities created what have been described as 'extended metropolitan regions', with intermingled urban and rural land use replicated around Jakarta, Kuala Lumpur, Bangkok and Singapore, in the latter case extending into both Malaysia and Indonesia. The same phenomenon arises to a lesser degree also around regional cities such as Bandung, Surabaya, Penang and Chiangmai (McGee and Greenberg 1992). It is in these areas that the interpenetration of urban and rural, identified by Rigg (2001, 2003) as a specifically Southeast Asian indicator of de-agrarianization, is most evident. Migration from villages to urban employment already created large urban populations by the 1950s. In the 1970s migration underwent a major increase, and for the first time began to involve a growing proportion of young women drawn not only from near-urban villages but also from more remote areas. Factory and other employment, albeit at low wages and under demanding working conditions, provided an attractive alternative to field and household work in the villages to young people of both sexes.

There is a huge literature on these migrations, and most of the discussion of contemporary change in Rigg (2003) is concerned, directly or indirectly, with modern rural–urban interaction and its consequences for people, their activities and the land. In regard to the modern movement of young women to factory employment, he approaches a conclusion at p. 275 in agreeing with Wolf (1992: 135) that 'although factory work organization and discipline are strict and often brutal, female workers perceive factory employment as a progressive change in their lives'. In short, they, and many young men also, prefer urban work to farming, with the consequence that agriculture in large parts of Southeast Asia has lost, or is losing, the most active part of its workforce. As was noted even in densely peopled central Java in the mid-1980s, there is significant de-intensification of farming: farm families, as collectives of their members, are 'too busy to farm' (Preston 1989).[4]

The distinction between city and country is becoming blurred. In addition to villagers commuting from village to city, there is new business in renting houses to migrants from further afield who work in the city, and buying of land by city people for investment or simply for speculation. Beyond the range of daily movement between home and workplace, there is a wide zone in which city-working villagers return home at weekends or less frequently. Rigg (2003) discusses the transformation of the countryside in Southeast Asia in considerable detail. Everywhere, even in remote areas, off-farm employment is of growing importance. Where agriculture continues to thrive it is on a fully commercial basis. Where non-farm work is available within easy reach, it is usually sought for preference, and the emergence of a wider range of rural industrial establishments, linked to the national and international markets, has widened the scope for diversification of household economies. So also has the growth of international migration, both within the southeast Asian region and beyond, to Japan and the Middle East in particular.

Households may continue to plan as aggregates, but individual members may follow very different trajectories, including long distance and even international migration as well as local and regional employment. Rigg sees this as a transition in progress. Viewing the evidence from a range of studies, he concludes that while rural people are reluctant to give up their security in land, the probability is that continuing structural transformation will lead to the situation where 'ultimately, most farmers will give up agriculture altogether' (Rigg 2003: 233). A return to Malaysia provides some corroboration of this view.

Two villages near Kuala Lumpur, Malaysia

Close to the Malaysian capital city, Brookfield *et al.* (1991) studied what we called the 'in-situ urbanization' of two rice and rubber villages in the mid–late 1980s. Beranang, 40 km south of Kuala Lumpur, had 3,000 people. In 1966, 49 per cent of the 805 ha mapped was under rice, and 30 per cent under rubber. By 1986, these percentages changed to 32 and 37 per cent respectively, by the loss of a good deal of rice and of rubber still being re-planted in the 1960s.[5] Beranang was a productive rice area, part of it growing two crops a year, but it had water problems, inadequately resolved by a new main irrigation canal to the largest wet-rice block, and the installation of higher level plastic flumes to carry water to the fields. There was a significant number of absentee landholders whose land lay idle once the tenants had gone to the city, and in 1989 we were told that the only clients for rented land were illegal Indonesian immigrants. Among the whole household population surveyed in 1986, 75 per cent of income was derived from non-local sources, and 95 per cent of it among the best-off 25 per cent of households. By 1989 much more rice land had become neglected and overgrown. An industrial estate and housing area only 2 km to the north was completed by 1989, with several factories already in operation. In the village, accommodation for tenants was built. The future of Beranang was clearly to be part of the urban complex. By the census year 2000, there were 10,500 people in Beranang and the industrial estate to its immediate north.

Janda Baik in the hills northeast of Kuala Lumpur was a much younger village, firmly established only in the 1930s on land earlier occupied by aboriginal people (*orang asli*). Its rice economy was the creation of a canalized irrigation scheme installed by government in the 1930s and its rubber mostly of post-1945 planting. In 1966, this relatively isolated community, surrounded on all sides by forest which occupied 62 per cent of its territory, had 10 per cent of its land under rice. Some 22 per cent was under rubber, almost half of it new, planted under a national scheme in a large block recently cleared from forest.

An expressway to the east coast, passing only 5 km from Janda Baik, was opened in 1980. New accessibility of this scenically pleasing upland area with cool nights began to attract weekend visitors and by 1986 there were 27 new houses built by or for such visitors, increasing to 40 by 1989. Market gardening, fruit and fish ponds took the place of some of the wet-rice land which declined to about half its 1966 extent by 1986. The market gardens supplied produce sold in a group of market stalls recently erected on the highway. About half the rubber, and the entire new high-yielding block planted in the 1960s, was neglected and became secondary forest. There was almost no tapping with low prices in 1986.

After prices improved in 1988, some untapped rubber was returned to production, but no rice was planted, and only a very small area in 1989. The irrigation ditches were stagnant and becoming overgrown. Even the market garden and other agricultural enterprises had fallen into disuse in 1989. Although there was no bus from the highway into the village, many villagers were working in Kuala Lumpur, including some of the leading men. In 1986, 55 per cent of household heads still regarded themselves as farmers, but 12 per cent were in skilled or semi-skilled non-farm occupations and 5 per cent in business. Although household incomes were lower in Janda Baik than in Beranang in 1986, the percentage from farm sources was substantially greater. Some farmers, who worked for and with Chinese farmers in the illegal but profitable cultivation in the forests of ginger for the city market, achieved better farm incomes than most farmers at Beranang. They included a number of the *orang asli*, grouped near the forest fringe at the eastern end of the community. In this village, unusually in Malaysia, the *orang asli* were in the middle-income group of households.

In 1989 Janda Baik saw the start of the next stage in transition. One farming enterprise had been converted into a rest and training camp for the ruling Malay party, and two small tourist establishments were under construction. The proposed building of a large hotel on 20 ha of valley and hill land was announced. By 2007 it became a simple matter to find Janda Baik on the Internet as a 'resort-and-residential' community. It had become a base for forest excursions by foreign tourists in four-wheel drive vehicles. This development is similar to the sort of new activity near Chiang Mai in Thailand described by Rigg and Ritchie (2002). As in Thailand, inflation of land values had accompanied these urbanizing activities. Using the records of land transfers in the land-title offices, we found that in the mid-1970s, median transfer values had been considerably lower at agriculturally poor Janda Baik than at Beranang, but developments had reversed this by the mid-1980s. Over 12 years, median transfer values, reduced to constant 1980 Malaysian dollars, increased by over seven times at Janda Baik but only by just over three times at Beranang.

China

It is rather remarkable that the de-agrarianization literature has not more fully taken up the most striking case of all, the massive growth of town and village enterprises and farm sideline activities in China since the 1970s, discussed in Chapter 7. Many town and rural township governments redefined themselves as industrial corporations, and all the more successful became employers of outside labour – even urban labour – in both the industrial and agricultural wings of their businesses. Teng Tou in Zhejiang is a good case, 27 km from the large city of Ningbo, south of Shanghai (Sanders 2000). It had 795 people in the late 1990s. It was a brigade in a larger commune during the collective period, and a strong leader was able to defy the party during the Cultural Revolution and plant fruit trees as well as just grain. There was more mechanization than in many collectives and the first factory was established in the 1970s. There were three by 1980, including a clothing factory which remained the largest and most successful. By 1997, 16 enterprises were run by the collective. They employed 1,300 workers many of whom were bussed in from other villages. It became a prize-winning 'ecological village' even though its farming is now wholly mechanized. It was better described as a 'thriving economic unit paying considerable attention to the quality of its natural environment' (Sanders 2000: 177). Disregarding what may have happened to Teng Tou since, with the failure of many of China's town and village enterprises, this would seem to describe it as a model of successful 'de-agrarianization'.

Japan

We presented a vignette of a farming community in Japan in the early twentieth century in Chapter 2, noting that the village described was not among the many that had participated in the widespread rural industrialization of the preceding century. Since that time almost all Japanese farming has become part-time farming, combined with employment or business either nearby or at a modest distance. The origins of this go back a long way. In Kaminoseki county, near the western end of the Inland Sea, the vast majority of farm households undertook some form of by-employment in the 1840s, providing from 20 to as much as 70 per cent of household income (T.C. Smith 1988). Even then, this was not unusual. Francks (1996, 2005) shows how mechanization geared to the small scale of Japanese farms and early rural electrification made possible the continuity of household livelihood diversification through the whole period of modern Japanese industrialization, involving both men and women on the farm, in the farmhouse, at nearby small factories or away in the cities. A growing proportion of the off-farm opportunities attracted male labour over time, drawing the women back to the farm to sustain the household base. By the late 1930s over half the farm labour force was female. On a basis which required only one household member to have off-farm work, over half of all farms were classed as part-time in 1938, then reaching 65 per cent before the end of the 1950s. By the early 1980s about 85 per cent of all Japanese farms were part-time, as today: 'Much agricultural work is carried out by women and the elderly, while younger men commute to off-farm jobs' (Francks 2005: 457).

This distinctive pattern has been backed up by a distinctive political economy. Large fields of industrial production have been organized to resolve problems of coordination and marketing among diffused small-scale producers who do not become proletarians. Governments have encouraged the rural dispersion of industry, while supporting agriculture with subsidies and very high tariffs, especially for rice. But while household diversification has yielded good results in terms of maintaining a prosperous rural population in many parts of Japan, it does not work so well in remote areas where the supporting infrastructure is weak and where there are no rewarding jobs for the young. There is now an elderly rural Japan that is much less than prosperous, and where not many of the farms are likely to survive beyond the lifetimes of their proprietors. Then there will be a more real de-agrarianization than anything else encountered in this chapter!

Notes

1   In South Africa, following limited enlargement after 1936, reserves occupied only 12.9 per cent of the national area, and were earlier yet smaller. This degree of territorial partitioning was designed to create African areas which could not become self-sufficient and would necessarily depend on labour migration to the cities and farms in non-African areas. In this country, de-agrarianization had thus been enforced. A government commission in the 1950s, known as the Tomlinson Commission after its chairman (South Africa 1955) was charged with proposing ways to make the reserves more self-sufficient as the basis for South Africa's peculiar variant of the world-wide drive for economic development – the 'apartheid' policy of separate development. By no means all the proposals were accepted, and few were adequately implemented. It was in any case a hopeless enterprise, as many critics remarked even in the days before the reserves were grouped into 'Bantustans'.

2   Tanzania ceased fertilizer subsidies to its farmers in 1994, leading to widespread concern that the national grain market would be insufficiently supplied, thus requiring increased and costly imports. In the early 2000s, subsidies were restored for maize farmers in the productive southern highlands, where Kinyanambo is located, but marketing to the nation as a whole from this one region is problematic in the absence of an adequate transport system (Kaihura 2005).

3   Jayne et al. (2005) differ in regarding some measure of land redistribution as necessary, given that the bottom 25 per cent (by land owned) of rural households are virtually landless.

4   Hardjono (1987) writing of the village of Sukahaji in the hills south of Bandung in West Java, found that employment in the factory in the nearby town of Majalaya was very important to the community that could no longer supply its needs from its own inten-sively used land, even with important diversification. Her main concern was that the small, old factories in Majalaya would be unable to compete with newer, better-capitalized industry, thus depriving the people of Sukahaji of employment that was essential to their livelihood.

5   The land use data for these two Malaysian villages in 1986 are from field survey. Data for 1966 are from interpretation and plotting from high resolution air photography taken in that year. All data were originally plotted on the 1:6,336 cadastral plans drawn up between 1900 and 1960.

13 Two paths into the new century
Pluriactivity and organics

The value of the long view

Having carried de-agrarianization as far as Japan, it is appropriate to see how useful the concept is in regard to modern Europe. Like Japan, but unlike Africa and much of Southeast Asia, Europe has a strong body of rural historiography, and this makes possible a long view. The two key elements of de-agrarianization, as discussed in Chapter 12 – supplementing agricultural incomes with earnings from other sources and rural out-migration – have been world-wide phenomena through most of the twentieth century. Yet the term de-agrarianization is rarely used to describe trends in developed countries, though we cite one of these rare instances below. Wilson and Rigg (2003) came closest in advocating that it be evaluated alongside currently used terms which we discuss later in this chapter. But an enthusiastic response is unlikely. The type of de-agrarianization reviewed in Africa and Southeast Asia happened in Europe at a time that is already history. In Chapter 6 we traced the rapid post-1945 decline in numbers of farmers and farm workers in Europe and the USA, a trend that had been going on since the late nineteenth century. Decline in farm numbers continues, but more slowly, and in most of Europe rural depopulation is now at an end.

Diversification of income sources is also far from new. In the unique early-modern English case, the non-agricultural proportion of the rural population rose from around 20 per cent in about 1520 to 50 per cent by 1800, much of it in small-scale industry, but a major share in commerce and services (Wrigley 1985). The non-agricultural share of rural employment continued to increase in the nineteenth century, reaching 66 per cent by 1900. In France, the non-agricultural members of a larger rural population were from 1500 to 1800 more numerous than in England, but the rate of growth was far less and its share reached 65 per cent only by 1950 (Collantes 2006b). How far the non-agricultural and agricultural rural populations were members of the same or different households is almost unknown, but it is likely that a proportion already represented the reorientation of farm household livelihoods and activities signalled for modern Africa by Bryceson (1996). In Europe, though, it would be termed pluriactivity.

Pluriactivity

Looking back assists in interpreting modern debates which have, until recently, been dominated by sectoral thinking: rural is agriculture; a real farm is a full-time farm; and if a farm household cannot make ends meet without seeking non-agricultural work it is likely to be on the way out of agriculture. In the UK, later than in much of continental Europe, it was not until the 1980s that it began to be accepted that 'part-time farming' was not necessarily any less efficient than full-time farming, and the term pluriactivity (borrowed from the French) was used (Fuller 1990; Shucksmith and Winter 1990). By including support derived from reciprocal arrangements, and also from welfare state transfers, pluriactivity embraces 'activities, . . . both on and off the farm, for which different kinds of remuneration are received (earnings, incomes in-kind, and transfers)' (Fuller 1990: 367). Nowadays, a majority of farm households in Europe and in many developing countries, and in North America too, can be described as 'pluriactive'.

Pluriactivity takes a variety of forms. In many countries, non-agricultural work has been done on the farm for centuries. Some farmers still make cheese, but selling milk, eggs or poultry has increasingly been constrained by health regulations and has become a specialized activity. Commerce and service trades for the rural population continued into recent times, but many a small rural shop has been squeezed out of business by competition from nearby urban supermarkets. Importantly, farm-based pluriactivity has involved working for other farmers, on a casual or more professional basis. From this emerged the farm contractor, perhaps the most successful of pluriactive enterprises, though many end up becoming specialists. Even seasonal migration for income-earning purposes has in the past been common in developed countries. Other forms of pluriactivity include processing of farm produce (own or for others), and the type of 'putting out' industrial production that has continued into recent times in Japan, but which in Europe largely faded away during the nineteenth century.

Rural industries serving agriculture, and manufacturing and repairing agricultural implements, were still important in the early twentieth century when rural manufac-turing entrepreneurs first developed the tractor, harvesting machinery and other innovations. Before the mid-twentieth century, these industries had concentrated into large factories most of which were located in urban areas. Rural specialists in infrastructure work, such as excavating farm dams, making and repairing farm roads and felling trees, continue, and the growth of handicraft production in some areas follows on from cottage industry of the past. Associated business, like accommodation for travellers, is also far from new. Although on-farm tourism has been much promoted in recent years in Europe, the uptake of this form of diversification has been relatively small. Farmers often take on new forms of business, but mainly in primary production. Viewing 'rural restructuring' with historical insight, Hoggart and Paniagua (2001a) query whether significant reconstruction has taken place.

The greatly increased availability of motor transport has led to a major expansion of pluriactivity, in the form of off-farm work. There is often a gender dimension.

Where the 'worker-peasant' combination has been an accepted norm for years, as in Germany, some other European countries and Japan, the male spouse often has the paid job, leaving main responsibility for the farm to his wife (Franklin 1969; Francks 1996, 2005). In other countries, it is more common for the female spouse to take off-farm work. In Norway the expansion of female off-farm employment has grown in parallel with female participation in the wage economy as a whole (Jervell 1999; see also Chapter 4). It is a part of a far more general social trend. Nonetheless, periods of downturn in rural economies are always times of hardship for farm families, who find it hard to make ends meet without recourse to off-farm employment.

From the mid-1970s and growing sharper in the 1990s, a contraction in real subsidies, an increase in the costs of farm inputs and a declining trend in market prices for farm produce produced a situation in which smaller-scale farmers suffered severe losses in effective income. Even in regions where off-farm pluriactivity was not common in the past, such as on the small livestock farms of Northern Ireland, it became an imperative (Shortall 2002). With men working long hours to keep the farm going, the most common pattern was for the women to take off-farm jobs, whether or not as a career. Often they became the main financial support not only of the household, but also of the farm. In a region where the net farm income already consisted almost entirely of EU subsidy payments, this situation presaged a bleak prospect. It is not difficult to see this as the enforced proletarianization of the small farm household via its (female) breadwinner.

In the Republic of Ireland, by contrast, it had been government policy since the 1940s to encourage location of small industry in rural areas, unlike the United Kingdom which followed the rural-preservationist Scott Report of 1942 in its 1947 Town and Country Planning Act (Tichelar 2004). As also in Denmark, Germany and Austria, there is a much stronger base of non-agricultural employment in rural areas, and pluriactivity is regarded as a normal and even satisfying way of family life; the surge at the end of the twentieth century merely accentuated a long-term trend (Kinsella *et al.* 2000). In the Republic of Ireland as a whole, 30 per cent of farm men and 23 per cent of farm women had off-farm jobs in 1998, and the share of off-farm income of total income rose from 19 per cent in 1973 to 40 per cent.[1] In central Ireland the proportion was almost 60 per cent (Kinsella *et al.* 2000). Money was an element in all family decisions to go pluriactive, but in County Clare, western Ireland, within commuting distance of the Shannon airport industrial area, the urgency to save the farm noted in Northern Ireland did not seem to be a major factor. Nor was it in most cases among the Norwegian farm households studied by Jervell (1999). We have to consider opportunities as well as constraints.

Off-farm employment is patchy, principally because opportunities are unevenly spread. Although some farm family members commute over surprising distances, the available opportunities depend on the nature and dynamism of the regional economy, as well as on rural planning policies. Except for work on other farms, or in forests, mining and fishing, most off-farm work is found in commerce and services, principally in towns both large and small. Industrial employment, notwithstanding trends toward dispersal especially in the last 25 years, is much more highly concentrated (except in Japan and Germany) and available only to long-term off-farm migrants. During the high-industrial period, which differed in time between European countries, this was the main pattern, and it accompanied major losses of rural populations just as in modern Southeast Asia (Chapter 12).

Pluriactivity and migration

Pluriactivity and migration have to be seen as complementary or alternative strategies for farm family members in periods of rapid change. Both strategies may be used if opportunities are present. Writing of rural northern Spain over the long historical span of the last two centuries, Collantes (2004, 2005a and b, 2006a) shows how farmers and their families adapted well to the changing conditions that arose from Spanish industrialization and urbanization until after 1950. A generalized livestock-based agriculture became commercial beef and dairy farming; off-farm work opportunities were seized, and some family members moved away to the cities. After liberalization of the national economy in 1959, emigration became absolutely dominant among young people, often with parental encouragement. Among a well-informed population, comparison of living standards between home and the cities, rather than of wages, encouraged massive migration, so that the regional rural population declined from 1.9 million in 1950 to less than 1.5 million in 1990.[2] Population decline in turn reduced the scope for non-agricultural activities, and hence opportunities for farm-based pluriactivity.

The consequence for rural change in most of rural Spain was that non-agricultural employment remained mainly in services to agriculture. When farm sons and daughters moved into the cities and the new tourist regions, as they did in great numbers in the 1960s and 1970s, the non-agricultural economy also shrank, but less massively. While the proportion of the whole rural population that was engaged in industry and services rose, it yielded only an apparent reconstruction of the rural economy that in fact represented very little real change (Hoggart and Paniagua 2001b; Collantes 2005b, 2006a). In most of the country, there were few rural jobs outside agriculture, and where they existed, they offered only low wages (Hoggart and Paniagua 2001b: 67). Entry into the EU facilitated further expansion of the olive, wine, fruit and vegetable production of southern Spain, but the dairy economy of the north contracted with international competition. Productivity in arable farming remained low by comparison with northern countries and, although farmers responded as far as they could to the loss of family workforces by mechanization, seasonal labour shortages were principally made up by hiring migrant labour coming (often illegally) from north and west Africa (Hoggart and Mendoza 1999).

Focusing more specifically on the 'less advantaged' mountain areas of Spain, Collantes (2004) found that a de-facto de-agrarianization (*desagrariazición*) had already taken place before the state embarked on a programme of support to the mountain farmers in 1981, in line with EU policy but ahead of Spanish entry into the EU in 1986. Agriculture was no longer the principal employment. Much the same was true of mountain areas of other European countries given support in these years, for example in the Massif Central of France from the 1960s, and in the Scottish highlands and uplands under the EU directive of 1975. The Scottish highlands were emptied by a much earlier depopulation produced by the infamous 'clearances' of the early nineteenth century, but in the Massif Central, the agricultural share of employment did not fall below 50 per cent before 1950, then declined to less than 25 per cent by the 1970s. As in some mountain areas of Spain, this late and rapid change was due to depopulation selective of the farmers among the total population.

Rural re-population

Within a very short time, many of the vacant farmhouses in the Cevennes, in the south of the Massif Central, were occupied by lifestyle migrants who had been students at the time of the 1968 counter-establishment 'uprising' in Paris and other cities. Some of these former squatters remained in the Cevennes, supplementing welfare incomes with the making of cheese and chestnut jam to sell to tourists, and with agri-environmental work restoring old farm buildings and walls (Willis and Campbell 2004). Other abandoned farmhouses are seasonally reoccupied by their former owners on holiday from their city employment. No similar wave of squatters resettled abandoned farmhouses in Spain, but Fernando Collantes (pers. comm. 24 October 2006) reports that many are retained as second homes, and others are converted for rural tourism. Compared with the rest of Europe, the 'counter-urban' movement in Spain is small.

Migration from the cities into the rural areas began a long time ago. Successful Londoners bought farms in the surrounding country in the late eighteenth century, though few managed to make a go of them (Middleton 1798). Right through the nineteenth century many of their successors bought or built rural houses, and in the twentieth century this became a substantial movement. From the mid-twentieth century onward, there was a fast-growing market for vacant farm houses, cottages and other farm buildings that could be converted into residences. Former labourers' houses were eagerly sought after and reconstructed by their new owners. Some of these turned out to be medieval buildings once stripped down. As the number of farms declined through purchase of land, abandonment or simply retirement, the market for rural houses expanded to provide a significant source of income for rural families. It became international as British, Dutch and Germans especially sought second homes in France, Italy and Spain, and more recently in southeastern Europe. In Spain, vacant houses in places with fewer than 2,000 inhabitants constituted 20 per cent of the national housing stock in 1990 (Hoggart and Paniagua 2001b). It was much the same in large parts of western and southwestern France, making it possible for purchasers with only limited resources or credit to buy farmhouses and village houses. Like the national urban migrants who had kept, or inherited, their former rural homes, the buyers mostly occupied them only seasonally and the associated land, if not sold or rented separately to neighbouring farmers, became overgrown.

Other ex-urban arrivals, mostly in regions closer to the cities and even quite small towns, have bought houses and land with the object of farming. Unless they have pensions or other secure income, most of them retain urban employment and farm only on a part-time basis. Contemptuously described in much of the literature as 'hobby farmers', many of these new arrivals take their farming very seriously, even though it is mostly on a relatively small scale. Some retain urban residences, and the wealthy among them even have their farms run by employed managers, but the majority live on their farms, which can be at a surprising distance from their urban employment. In effect, they have joined the population of pluriactive farmers, although in social terms they are rarely integrated into the rural community. Together with rural residents who do not farm, they are now a major element in the population surrounding cities, not only in Europe, but also in North America and Australia. Indeed, one of us is such a part-time farmer.

The effect of modern in-migration in rural areas has been hotly debated, especially in relation to housing. The incomers are able to outbid local home-seekers and, where the housing stock is limited, this can lead to significant tensions. Those who retain urban jobs, and commute, generally find rural shops inadequate and expensive, and buy their supplies from supermarkets in town, further reducing the competitive position of the rural stores. Some create jobs, but others, who are better qualified than local people, often get the few available jobs. In a wide-ranging study of rural in-migrants in Scotland, Stockdale *et al.* (2000) found on balance that local rural economies benefited from the new migration and its infusion of spending, but that there has been only limited rural revitalization. Where the in-migrants are more numerous, as they are close to large cities and in many scenically attractive areas, they can change the nature of rural society and economy in a comprehensive manner.

The incomers, including those who set up family farms, tend to have urban values on environmental management, and are 'consumers' of a countryside that has been created and maintained by generations of farmers, latterly under policies that have encouraged and even enforced the modernization of their practices. Buying houses and land, the incomers are replacing, or settling among, the smaller-scale family farmers who have been increasingly squeezed by rising costs and falling prices. They are moving into a countryside in which agriculture is no longer the principal rural occupation and are only one element in what is nowadays called the 'consumption countryside', which much of the European population, and large numbers of visitors, value for its created environment (Marsden 1999). They are the most obvious reminder of the growing diversity of rural spaces and of their functions in European societies and economies.

How social science has viewed these changes

Before we turn to the most distinctive of practical responses of farmers to the new conditions, we review how they have been seen by concerned outsiders, the social scientists involved with rural issues and the bureaucrats in Brussels and national capitals. Both groups have been very slow to see rural problems in terms wider than farming. The large regional science literature that accumulated after 1950 has had a strongly urban bias, while the problems of equalizing developmental levels within the EU and within countries were seen in terms of facilitating the wider distribution of industry, enhanced infrastructure and commerce. Not until the 1990s did it become clear to all that the goal of sustaining a large farm population in the EU could not be met by farm-support measures alone, especially as this support was so unequally distributed. First in rhetoric, then in more practical terms, the aims of policy shifted away from supporting production toward the newer emphasis on environmental management and regional development (Chapter 10).

'Post-productivism'

In social science, this was the era of 'post-modernism', and soon also of 'post-Fordism', 'post-development' and, thence, unsurprisingly, 'post-productivism'. The latter term first appeared in papers written about 1992/3, but the ideas were earlier (Wilson 2001). In a literature virtually confined to the UK, although with abundant reference to Europe, it was argued that the main elements of 'productivism' (intensification, concentration of land, specialization of production) had been reversed, an argument empirically rejected by writers such as Evans *et al.* (2002). The term 'post-productivism' was widely, even uncritically, adopted in the UK literature but, though tried and found wanting in Australia, was taken up nowhere else. It related particularly to the policy environment, especially that part of it concerned with agri-environmental issues, but the implication was of a major change of heart among British and European farmers. Three good empirical papers, concerned mainly with larger family farmers, showed that there was no such change of heart in English farming (Walford 2003; Burton 2004; Burton and Walford 2005). Farmers continued to be concerned with the results of their own productive efforts, while judging neighbours on the state of their fields and crops, much as in 1930s Nigeria (Chapter 2). Many of these farmers met set-aside requirements from their least capable land, intensifying inputs on the remainder.

Re-analysing data from research among English Midland farmers in the 1990s, Burton and Wilson (2006) emphasized that many self-identified 'good farmers' are convinced that their productivist ('agricultural producer') primary goals are not incompatible with conservation, whether or not environmentalists would agree with them. In the construction of their self-identities there may therefore be a small but perceptible shift in the direction in which policy makers, academics and many of the public seem to want them to travel. The shift is small, and for farmers in any part of Europe to adopt primary identities as environmental managers and producers of 'consumption space' for non-agriculture would demand that both farmers and the public change thousand-year-old notions of what farming is about (Burton and Wilson 2006). Such a shift is more likely where farming is no longer profitable, and where the new agri-environmental subsidies have become significant, perhaps leading, as Wilson (2001) earlier suggested, to a differentiation of European agriculture into 'productivist' and 'post-productivist' spaces. More dramatically Bové and Dufour (2005) unenthusiastically envisaged a future in which French agriculture would divide between regions of intensive production (like the Paris basin) and regions where farmers are park keepers without uniforms. Either way, new rural spaces are being differentiated by current trends.

Multifunctionality and 'rural development'

One approach that seems to be overtaking mythical 'post-productivism' came, in the first place, not from academic writers but from the corridors of the EU in Brussels, where a new formula was being sought in the 1990s to justify European subsidies in the face of rising criticism within the WTO. The justification found, and subsequently strongly pressed, was in the 'multifunctionality' of farming, not of the farmers themselves. Wilson (2001) suggested that there is a transition from all-out productivism to multifunctionality rather than to anything else. Wilson, it would seem, meant that agriculture supplies environmental goods as well as food and fibre. We could add that it also supplies labour to a range of non-agricultural activities in both rural and urban locations. But this systems approach draws attention away from the actors. As a descriptor of land use, multifunctionality is trite. Farming of land has always served purposes additional to the production of food. As a trend, it can be seen both as part of the localizing phase of modern neo-liberalism, and as opposition to these politically guided trends (McCarthy 2005). On its own it would seem to get us little further forward, though it is without the misleading implications of 'post-productivism'.[3]

One important approach has arisen from work focused at Wageningen in the Netherlands, with participants in most European countries, including Britain. Beginning with a focus on the range of farming styles in Europe, and on farming economically, the confusingly named 'rural development' approach has evolved through a coordinated group of studies of local initiatives across Europe. These have in common the object of escaping the standardizing trends of the modernizing, or productivist, project of the last half-century (van der Ploeg *et al.* 2000; van der Ploeg and Renting 2000). Thirty very different cases yielded evidence of farmer enterprise in new or higher-quality products, success in marketing through more direct channels, new enterprises, pluriactivity and, in most cases, substantial income improvements. Farmers were using on-farm resources of labour, buildings, equipment and skills in new ways, usually replacing purchased inputs by on-farm recycling, and exploiting 'economies of scope' through multiple use of their resources, rather than economies of scale. In such ways, multifunctionality became a means to an end.

Van der Ploeg and Renting felt they had discerned a new paradigm of farm-based rural development, with significant benefits to whole regional populations. In a later paper, responding to a critique which we discuss later in this section, they argued that farmers were 'deepening' their activities by going in for quality or regionally distinct products, or going organic, and developing short supply chains to their customers. Such deepening alone was calculated, or perhaps estimated, to provide additional added value equivalent to the total value added in the whole agricultural sector in the Netherlands. Many were also 'broadening' by diversifying production, undertaking new on-farm activities or going into nature and landscape management, while some were 'regrounding' the farm base through off-farm income and devising new forms of cost reduction (van der Ploeg and Renting 2004).

A widespread characteristic of the new practices is far stronger emphasis on the quality of farm products than under the standardized methods of productivism. The aim is to escape the 'technological treadmill' of continual innovation to reduce costs in order to produce for the demands of the increasingly concentrated processing, packing and retailing system (Chapter 5). Such producers are responding to what is called the 'quality turn' in consumer preferences, which has grown strongly since the 1990s in response to growing disquiet about the safety of some food. In terms of a large theoretical discussion around which we are deliberately skirting, these producers are seeking to move away from a standardized-generic production system to a more specialized system without, in most cases, becoming dedicated to such a narrow market that possibilities of expansion are shut off (Murdoch *et al.* 2000).[4] Success is favoured where there is an internationally recognized region-based specialized production, of which the *appellation d'origine contrôlée* is the best known. These legal designations, now approved and reinforced by the CAP, are most widespread in France and Italy; they are much less significant in northern Europe.

One example is the modern recovery of artisanally made Parmigiano Reggiano cheese, which has been defended against stiff competition by its legal name protection but still has had to reduce costs to maintain market share (de Roest and Menghi 2000). Like other examples, this required farmers' cooperation over a defined region. Because some capital is required to support regionally distinct quality production and its marketing, the cooperating farmers are generally not the smallest. Another strong example arose from the establishment of a large organic dairy in a German biosphere reserve, encouraging farmers to undertake organic milk production. In this case a number of other activities, all 'consuming' the countryside, were added to milk production, thus providing regional multiplier effects (Knickel and Renting 2000). Both these examples relied on good marketing. The biggest problem is being in direct competition with conventional producers' supply chains to the major supermarkets. This is especially so where the 'rural development' farmers are seeking premium prices for their distinctive produce. This was the problem with efforts to revive bread-wheat production in the southwestern Netherlands using organic cultivation and artisanal milling, in competition with conventionally grown bread-wheat imported from France and Germany. The quality of the bread was excellent, but no self-sustaining market could be developed, even within the small province of Zeeland (Wiskerke 2003).

A major set of objections to the arguments of the Wageningen school came from the still-Marxian pen of veteran rural sociologist David Goodman (2004). Like us, Goodman does not like the inclusion of widespread and established practices such as pluriactivity and 'farming economically', as well as funded nature and landscape management, within the broadly defined new paradigm. He is particularly concerned with the weight given to 'short food supply chains' (Marsden *et al.* 2000; Renting *et al.* 2003). These short chains, short in number of intermediaries as much as in geographical length, must be lengthened where larger markets are sought. The economic rents obtained from quality products, erodible by competition, are also erodible by effective incorporation into the longer chains developed by the supermarkets and traders. This leads to cost-cutting pressure on family workforces which, Goodman argues, the Wageningen group disregard. Beyond this, regional quality products are purchased mainly by the affluent, so that, for the moment, these innovative modes of provisioning represent 'socially exclusive niches rather than the future of European rural economy and society' (Goodman 2004: 13). Following Allen (1999), he also makes the valid point that the commercial system has had its own 'quality turn' while at the same time democratizing access to good food. In their enthusiastic search for a new paradigm to replace all-out productivism, the Wageningen group may have taken inadequate notice of the fact that their innovators are a minority among farmers, supplying a minority among Europe's customers. This is an interesting debate, though it is not comprehensive as Box 13.1 demonstrates. We now look more closely at organic farming, one of the more widespread innovations supported by the EU. This also takes us outside Europe again.

Organic farming

Pluriactivity is very common; organic farming is only followed by a minority and unlikely ever to grow far beyond 10–15 per cent of Europe's farmers. Nonetheless, this well-defined and well written-up segment of modern farming can stand for a much larger movement away from industrial farming among the family farmers of developed countries. We do not seek to argue that it is the answer to all of the world's problems. We note from Smil (2001: 50) the fact that synthetic nitrogen production world-wide now exceeds the total amount of nitrogen received by farmland from atmospheric deposition, recycling of organic wastes and bio-fixation, and that elimination of industrially fixed nitrogen would reduce the world's capacity to support people by several billion. To produce the present volume of world food by organic means would, in any case, require very much more agricultural land. We also note that in our own country, organic production is so much considered non-viable by a national research organization concerned with plant industry that they will put no money into it. Nonetheless, there is a demand for organic food, and it continues to increase.

Although modern variants of organic crop and livestock farming, rejecting chemicals and placing emphasis on practices that sustain soil fertility, began before World War II, they remained tiny minority activities until the 1980s.[5] Even in 1985, organic production in the EU was confined to only 6,300 holdings, less than 0.1 per cent of the total number of farms. By the end of 1999, 127,000 farmers were producing organically, on nearly 1.5 per cent of holdings and 2.4 per cent of the total agricultural area (Padel 2001: 42). Rapid but very uneven growth took off in the mid-1990s and has continued. There was parallel expansion under somewhat different conditions in North America and the Antipodes, and there has been significant growth in China (Sanders 2006). Conviction-driven movements like 'Fair Trade' made organic production known to many developing country producers of tea and coffee, and of bananas as well. Some of them may not fully understand how 'Fair Trade' arrives at its prices, supposedly based on production costs, but they all understand the meaning of 'organic' (Shreck 2005).

In Europe, organic farming has been considered an environmentally friendly method since the 1980s, and gained EU support in the 1992 reform of the CAP.

BOX 13.1 FROM CAPITALIST TO FAMILY FARMING IN GERMANY

Germany was a land of relatively small farms before the 1950s, and anything over 50 ha was regarded as a 'capitalist' farm (Franklin 1969: 27). One leasehold farm, in the northwest, still operates 75 ha today, and was much larger before the late 1940s when a large part of its land was resumed to settle refugees from the eastern regions lost to Poland. At its peak it had employed 50 regular workers in mixed arable-livestock operation. Numbers of labourers declined rapidly in the 1950s as the farm mechanized, but some of the live-in workers were kept on despite their lack of mechanical skills when they could not easily find other employment. The first son took over the farm when his father died, but was attracted to other employment, and the third son, an engineer, then became the leaseholder. He and his agriculturally trained wife decided to specialize in potato production and poultry, egg and chicken meat as well as seasonal geese. Poultry manure is composted and returned to the fields. They have built new housing for the poultry to automate and reduce labour inputs, and to comply with regulations for animal slaughter and egg packing. Only casual farm labourers now remain, although the now adult children still help out. The farmer and his wife do the major part of the work on the 3,000-ton crop of potatoes and with the poultry, even hands on in the poultry slaughter. The farmer recently has designed and built a potato storage facility that uses cold night air to keep temperatures down to 13 degrees and minimize use of electricity. They are able to sell on spot prices just before the next season, and avoid having to contract.

The farm is on the edge of a village with small towns within 10 km and two cities 40 km away. Soon after starting to grow potatoes and keep chickens, they opened a small shop to sell some produce direct to consumers. They began local direct marketing to shops and bakeries several times a week in the villages and nearby cities, delivering using their own mini-van, and sometimes casual labour. The shop now also sells the produce of neighbours, locally processed foods and other stock-in-trade to meet the village demand, since other village shops have closed. Two Russian women work full-time in the shop. The whole thrust has been to mechanize and automate in order to save labour, and to economize on external inputs and services. Technical training and skills gained before farming have been invaluable in developing the new ventures. They rent some potato storage space to other farmers and advertise their shop through a website, in addition to running a 'farm festival' once a year when city and town people can visit the farm. In these ways they have made a successful business out of a farm of modest size (Gisela Pullen, pers. comm. 2007).

Since 1999, the EU as a whole has subsidized farmers for the inevitable income losses during the conversion process. The subsidies have assisted many farmers to decide to convert, but there are also other older and still persuasive reasons. One is a wish to work the farm as a substantially closed system with regard to organic matter and nutrients, managing weed growth without herbicides, and controlling pests and diseases by biological methods.[6] The aim, as it has been from the beginning of organics as a social movement in the 1920s, is to sustain soil health while producing healthy food. This is powerfully reinforced by the wider drive to reduce the cost of external inputs as their prices soared, while farm output prices have been consistently weak. As in North America (Committee 1989), a minority of farmers has successfully demonstrated that these goals can be achieved and profits made. A more direct incentive to convert is provided by the premium prices that organic produce commands, at least until now, because of public demand for food perceived to be healthy, and produced in ways that respect nature conservation and animal welfare. This demand grew rapidly through the food scares of the 1990s, and ceased to be confined to a small minority of affluent consumers.[7] Each country has developed its own detailed standards of certification, although a general EU standard of certification was determined in 1991.

For farmers, there are both advantages and problems in converting to organic. Other than the health advantage of escaping the need to work with chemicals, the principal attraction is regaining decision-making independence, or at least a large part of it. The problems are considerable, and they are reflected in the somewhat chequered growth of organic production in different countries. On the farm itself, a good deal of work that for 50 years has been done with chemicals now has to be done by hand or machine. Crop rotations and the cultivation of green manures are necessary. More seriously, there is almost always a drop in yield during conversion, and even after it is complete. There is consequently a drop in income which can be substantial, more than is compensated by subsidy. Certification of produce, or livestock, is only possible once conversion is complete, so any price advantage is deferred beyond the most costly period.[8] The certification process itself has considerable transaction costs, as well as requiring inspections that are certainly demanding. Moreover, not everyone is entirely convinced by the ecological arguments in favour of organic methods of production (Rigby and Cáceres 2000).

Characteristically, organics are first adopted by a minority of enthusiasts, then by a larger number who see advantages over conventional methods, reaching ultimately a few per cent of the farming population (Padel 2001). The majority among those who ultimately convert are pragmatists, not ideologists, and they include a share of environmentally conscious farmers who adopt organic methods without the commitment of signing long-term contracts needed for subsidies (Darnhofer *et al.* 2005). The main surge in Europe followed underwriting of a few earlier national subsidies by the EU beginning in 1992, then becoming more general as part of agri-environmental initiatives in 1999. The rate of growth then slowed almost everywhere. Adoption of the halfway house of 'integrated farming' or 'integrated crop management', using few external inputs and adopting crop rotation and other methods of organic farmers, has expanded much more but because it has no mandatory certification or subsidy, it is inadequately recorded (Morris and Winter 1999).

In the early days, organic food was the archetypical example of the short supply chain, being sold either directly to consumers or to specialized natural food shops, keeping right outside the regular commercial system. It stood in opposition not only to mainstream farming but also to mainstream retailing. Organic farmers, along with other 'quality' producers, have been active in the farmers' markets that have sprung up almost all over the developed countries since the early 1990s – although something like them had developed earlier in Russia and other collectivized countries where such markets were a main outlet for produce grown on the individual plots of the collective and state-farm members. Already in the 1980s, and more widely by the early 1990s, supermarkets began to sell organic produce to satisfy a strong demand using long-distance supply chains. In Germany and the UK, a large amount of organic fruit and vegetables is now imported, mainly from southern Europe, but even from as far as New Zealand. As a consequence, while Italy has an unusually high proportion of organic farmers, consumption of organic produce in Italy is relatively small; production is largely for export to northern Europe.

Smith and Marsden (2004) found some trends in the marketing of organic produce that were disturbing to those seeing a bright future for the organic farmer in the UK and elsewhere in northern Europe. As organic milk production rose since the mid-1990s, there was a steady decline in the farm-gate prices received. Organic meat sold to supermarkets was often placed on the shelves alongside other 'quality' meats that were not organic. Organic canned or packaged foods are mixed with others on the shelves, leaving the customer to select organic produce if so desired. Even while continuing to proclaim support for the British organic farmer, and seeking more homegrown supplies, the big supermarket companies are both importing lower-priced produce from abroad and simultaneously narrowing the price premium paid to UK farmers for organic produce. The aim is to satisfy the consumer market for quality produce at prices no more than those paid for conventionally grown produce. That is, they are treating organic produce as just one class of commodity in parallel with others. Once organic farmers have entered the commodity chains that supply the supermarkets, they have few defences against this strategy. The farm-gate price squeeze that had earlier afflicted conventional UK farmers now began to be felt by the organic farmers as well, with no subsidy once the conversion period was over. Nonetheless, the demand for organic produce continues to grow.

The significance of the organic farming debate

The future of organic farming has been questioned on other grounds. It is a form of production particularly suited to family farm operation where, in the terms discussed in Chapter 1, it corresponds with 'simple commodity production', even though it is far from simple. Flexibility is a necessary characteristic of fully organic production, limited to biological responses to biological hazards (Coombs and

Campbell 1998). Especially in countries with no subsidies, organic farms also need to be sufficiently flexible to cope with the substantial costs arising from conversion and the need to establish and sustain a market. Once a market for organic produce is demonstrated, there will be competition from better resourced producers and traders. This happened dramatically in California, where large-scale producers entered the organic field as soon as a nationwide demand for 'clean and green' produce was first demonstrated in the early 1990s. This was possible because while organic production is unique in necessitating legal definition, such definition has not depended on science. Unlike the innovations of modernist (or productivist) agriculture, alternative agriculture, of which organic production is a central part, arose from among a minority of producers in cooperation with a minority of consumers. The science of sustainable agriculture has followed rather than preceded the innovators (Altieri 1995; Gleissman 1998). Government regulation, though late, also had little benefit from science. It was therefore possible for commercial interests wishing to enter a promising field on a large scale, and requiring the protection of regulation in order to secure premium prices, to propose a minimalist view of organic production, confined almost wholly to permitted and prohibited inputs, and ignoring all the ecological aspects of organic field methods (Rosset and Altieri 1997).

This is what happened in the USA, even though proposals to permit use of genetically modified germplasm, sewage sludge and irradiation were ultimately overturned.[9] It made possible an 'organic lite' (Guthman 2004), managed by industrial methods, and on only a part of any whole farm, selling to what became a powerful group of grower-shippers who supplied the whole country with 57 per cent of its organic vegetables (above all, lettuce) and 52 per cent of its certified fruit. Buck *et al.* (1997), in a paper of which Guthman was one author, argued that this conventionalization of marketing would, by out-competing other forms of marketing and creating the farm-gate price squeeze noted later for the UK by Smith and Marsden (2004), make adherence to full organic standards uneconomic for all but marginal members of a farm population. In California, this population did not spring from classic family farming, but operated on a capitalist basis with wage or contract labour. With both organic and conventional production operating almost wholly on irrigated land, high land values were further enhanced by writing the expected incomes from intensive production into land prices. A few years later, Guthman (2004) conceded that a 'vibrant' small-scale organic sector, selling only locally, had persisted in California, but otherwise confirmed that most producers had been 'conventionalized' much as had been proposed over half a decade earlier. A regulatory regime which does no more than support a price premium only contributes to the erosion of true organic practices.

Many have disagreed with the view that the intrusion of agribusiness necessarily spells doom to the organic industry. The Canadian province of Ontario, which has given minimal support to its organic farmers, in the 1990s was actively promoting conventional farming as being 'sustainable' (Hall 2003). Hall and Mogyorody (2001) found that small organic producers were experiencing little of the competition and pressure encountered in California. In stridently neo-liberal New Zealand, where a locally based organic sector emerged in the 1970s, large companies seeking organic produce for export found supplies by persuading conventional farmers to convert. The more diversified home-market sector not only survived but did so in useful technical collaboration with the exporting agribusiness firms (Coombs and Campbell 1998; Campbell and Liepins 2001). The firms, concerned about their image in the importing countries, especially quality-sensitive Japan, accepted the standards generated by the industry itself through its organization, Biogro. Later, the home-market growers felt that these standards were inadequate to cope with their diverse production systems and many withdrew to deal with their customers on a basis of trust. The ultimate outcome was two sets of standards, developed under belated government intervention.

The New Zealand case rather sharply underlines the mixed motivations involved in the organic movement, including its customers. Those producers and consumers who believe in the conjoined qualities of good food with biological cultivation and rearing, were the origin of the movement and remain its core. Other producers, who seek to increase their gross margins by eliminating the cost of inputs and seeking higher sale prices to compensate for their extra efforts, represent a much larger group and they are paired with a larger consuming population who might prefer organic produce but will not, or cannot, pay more than a small premium for it. Buyers of food brought from a distance, or imported, have to take its organic nature on trust, hence the importance of reliable certification.

But is this all? Are organics just another commodity in competition with others produced on different principles? Once local and nearby markets are saturated, and organic produce has to enter the commercial supply chain system, can organic farmers still compete, except in the marginal manner that some do in California? There are different views on this question, few of them optimistic (see Banks and Marsden 2001). One example is from lower Austria, a country with one of the highest relative shares of organic production in Europe, in a district where 11 per cent of farmers are organic. Ika Darnhofer (2005) worked in depth with a small number of both conventional and organic farmers, the type of inquiry that cannot provide statistically validated conclusions but, as we have seen in other places, can yield important insights. In a region where most produce goes through the commercial chain to a powerful oligopsony of supermarkets, a significant price premium is available only on the minority of sales through local outlets. What do farmers do under these conditions, and of what help is conversion to organics?

The Austrian organic farm households described by Darnhofer (2005) routinely practised pluriactivity. They commonly undertook contract harvesting for part-time farmers, using their own machines, or took off-farm employment. Some distilled brandy, raised geese, cut and sold firewood, or reared piglets, undertook agri-environmental work, or earned income from local government by composting recyclable urban waste, mowing the roadsides and clearing the roads of snow in winter. Organic farmers have set up biogas plants, using leguminous crops grown in rotation to yield electricity for sale and de-gassed biomass for their own use as fertilizer. In this way they can shorten rotations. Potatoes grown in rotation can be fed to pigs. By diversifying activities in these and other ways, they take advantage of the fact that with big savings on input, gross margins are higher so that enterprises can be profitable at a smaller scale than in conventional farming. Enjoying economies of scope, they are able to devote a part of their resources to non-agricultural as well as agricultural enterprises, expanding their income base, avoiding risk and increasing the quality of life and work. 'Farmers do not limit themselves to the food production aspects that dominate the organic farming debate' (Darnhofer 2005: 319). They deliver a range of products and services. In short, they survive by profitably 'de-agrarianizing' their activities and their farms. It is the flexibility that makes this possible.

Conclusion

Darnhofer's (2005) material is set in the mould of the Wageningen school regional development paradigm. As she puts it, adoption of organic farming is much more than simply seeking 'rents' from premium prices for quality or ecologically grown produce, as Goodman (2004) suggested. Whether, together with *appellation contrôlée* and other quality production, it represents a wholly new approach for European (and other) farming is far less certain. The German example in Box 13.1 demonstrates a very different path, though it uses some of the same devices. The need to escape the cost-price squeeze and technological treadmill of modernist production is very evident. But so also are the dangers of the technological alternative for reducing inputs offered by buying genetically modified seed, with its barely hidden purpose of tying the farmers inescapably to the seed-chemical companies (Brookfield 2001). There are several different levels of 'alternative agriculture'. They appeal to many farmers, although the consuming public does not necessarily distinguish them. Thus 'country fresh' is to many a more appealing label than 'organic'. The consuming public is a force of increasing importance in agricultural policy and we explore its role further in our final chapter.

Notes

1 In the turn of century period, Northern Ireland suffered serious losses of Euro-nominated income because of the high exchange rate of the British pound, while the Republic of Ireland had entered the Euro zone. For the UK as a whole, the (Curry) Policy Commission (2002: 24) remarked 'British farming is really a euro-area industry operating in the wrong currency'.
2 Collantes is applying the theoretical argument presented in Chapter 1. He is not comparing rural with urban standards purely in monetary terms, but also including quality-of-life intangibles such as leisure time activities, working conditions, housing and availability of consumer goods and services.
3 Although, to our regret, the term 'post-productivism' is not yet dead, it has been usefully redefined by Mather *et al.* (2006). Incorporating forestry as well as agriculture, these authors still find merit in the term in the analysis of changes in land use.
4 The specific areas of theory which we are skirting around in this chapter include 'ecological modernization', actor-network theory and convention theory. All have a range much wider than agriculture, but all have been drawn on in discussion of modern change. We are consciously avoiding 'ecological modernization' as advocated by Evans *et al.* (2002), find actor-network theory much more relevant since it incorporates natural elements and forces as actors, as necessary in all agriculture (but for that reason

also trite), and come closest to convention theory in broadly adopting the 'production worlds' of Salais and Storper (1992).

5 In much of Europe, organic farming is described, perhaps more aptly, as 'ecological' or 'biological' farming, the sense of 'organic' being of the farm as an organism, rather than of the nature of the inputs used (Padel 2001). In certification, the nature of the inputs has dominated, especially in the USA.

6 We have discussed the ecological aspects of organic production in some detail in our well-reviewed, but little-cited, earlier book (Brookfield 2001: 253–60; 270–80). Space does not allow a summary here.

7 It was particularly effective in the sudden increase in demand for organic milk, following the US Department of Agriculture's (USDA) approval of milk-yield enhancing hormone in cattle feed in 1993. In this case, the boards of large dairy companies exploited a market for milk that could be guaranteed free of this hormone, although they only reached 2 per cent of the US market by 1999 (DuPuis 2000). Taken in conjunction with the other food scares of the 1990s, this produced a contagious reaction among 'reflexive' consumers in other countries, so that organic milk achieved a rising market share also in Britain, for example, where the particular hormone was never introduced (Smith and Marsden 2004).

8 Commonly, land has to be worked according to organic rules for three years or more before full certification can be obtained. Meantime, any price premium available cannot be obtained (except by trust or collusion between seller and buyer).

9 Drawing on a wider literature, we earlier noted a group of commercial organic farmers in the USA whose version of organic farming includes use of 'a range of mineral and organic products such as seaweed, fish meal, crusher dust, potash, bone meal, bat guano, and commercially-produced compost . . . manufactured from widely separated sources and transported over long distances' (Brookfield 2001: 255–6).

14 Prospect
The coming generation

In this final chapter we try to look forward through one generation from now, let's say through 27 years hence to 2034. This may not seem far ahead, but when we consider all that has happened in the world of farmers during the last 27 years, between 1980 and 2007, it seems a dangerously long way. The justification for making the attempt at all is that there seems no present likelihood of a return to the political pluralism of the twentieth century, so that shades of neo-liberalism may remain the only set of available policies. We may or may not be right in believing that the technological changes in farming in the coming generation will build on those that have evolved in the twentieth century, rather than being of a revolutionary nature. We have to assume that the present headlong drive to produce energy crops will not, in some major countries, have so constrained food production as drastically to change marketing conditions for farming as a whole, and for smaller farmers in particular. We cannot take account of wars, but can at least hope they will be less savage than the localized horrors of the last generation. Neither do we discuss the anthropogenic climatic changes that are already evident, and are likely to present much more serious problems before 2034.

We also have to assume that the rising economic power of China, India and Brazil (and perhaps Russia as well) during the coming generation will have reduced the political weight of the North Atlantic countries but without yet eliminating their influence. On this basis, it is likely that there will be continuing pressure to eliminate subsidies and reduce tariffs, but also that these pressures will encounter stiff resistance. If the WTO survives, it cannot continue to be so totally dominated by North Atlantic interests. Our approach is to project tentatively forward from current political and social trends in the context of all we have discussed in this book, developing some further discussion of certain present trends that seem most relevant to the future of family farming. That is to say, we view the coming years as a projection of trends, and contradictions too, that are already established. In all this, we recognize that we are setting ourselves up as a highly vulnerable target for criticism, especially from the standpoint of 2034.

Visions for agriculture

To begin, we note the large body of shallowly economistic writers, and some influential journalists too, who maintain almost daily that agriculture is no longer of importance. Mindless of whence food is derived, they see only the small and declining share of agriculture in GDP and total employment in the developed countries, and disregard the large economic sectors that require agriculture as market for inputs or suppliers of produce for processing and retail. There are others, rather better informed, who regard agriculture as but one industry among others, and see no good reason why, to take one example, EU tariff protection for agricultural goods should average 20 per cent while for the products of other industries it now averages only 4 per cent (UK Treasury and Defra 2005). All these writers tend to be hostile to subsidies for agriculture. Drawing on the central ideas of a commission on the future of English agriculture (Policy Commission 2002), as well as on the neo-liberal principles that have ruled in Britain for more than two decades, the official British vision of European agriculture after about 2020 is of:

> an industry which is fundamentally sustainable and an integral part of the European economy. It should be [*inter alia*]: internationally competitive without subsidy or protection; rewarded by the market for its outputs, not least safe and good quality food, and by the taxpayer only for producing societal benefits that the market cannot deliver.
>
> (UK Treasury and Defra 2005: 9)

There are many, perhaps even in North America too, who would agree. Although subsidization persists on both sides of the Atlantic, it is far less secure than it was during the twentieth century. Some of it, but not all of it, will probably last through at least the first half of our short time span.

The British vision does not ignore developing country farmers, but treats them only in an international context. They should be able to export competitively, while not suffering the competition of imported produce that is subsidized at source. Again, there are many who would agree. Other writers are less concerned with competition and international trade, and put more emphasis on the challenge posed by the continued growth of developing country populations. They are worried about the possibilities of feeding a world population of 10 billion people, which we will almost certainly have by the middle of this century. We cannot discuss here the large and conflicting literature on this topic produced by the catastrophists and cornucopianists. Nor in this book can we go into the claims that corporation-controlled biotechnology offers the only way out of the problem, other than by saying that the corporate claims for much higher yields through genetic modification have still to be substantiated.[1]

We take these conditions of rising demand and sharpening constraints as inescapable context. With Vaclav Smil (2001), we see both the necessity and possibility of enormous improvements in the technical efficiency of agricultural production within the half-century time span which Smil adopts for his survey of future food-production prospects. His arguments concerning what is possible by improvements in the technical efficiency of, for example, fertilizer use and the use of water are very persuasive. Within our shorter period, we see no reason to suppose that the technical challenges cannot be met and, therefore, that while the ability to purchase food may grow worse for some people, there will be no global famine. However, the need for more efficient agriculture has substantial implications for its organization even within this limited time span, and we discuss these implications.

The conditions of efficiency

The view espoused by most national authorities, by the World Bank and by most afficionados of neo-liberal economic doctrines is that all farms must compete, and to do so they have to be economically efficient. This, in turn, means that they must adopt the most modern technology and reap all possible economies of scale. They have to build on the programme of the productivist era, intensifying the use of machinery, adopting the products of biotechnology and continuing to use chemical inputs on a large scale, but with greater technical efficiency and much more carefully in order to avoid the damage done in the recent past. The most sophisticated form of this vision – locally already a reality – has 'precision farming' managed from giant tractors with on-board computers and GPS guidance, working land of which the water and fertilizer needs are known in detail and applied accordingly. The knowledge and resources to perform all operations in the timeliest manner is essential. In this vision, there seems to be little place for the small family farmer. Governments should, it is therefore argued, support the acquisition of small farms by large farms through facilitating an efficient land market and perhaps easing small farmers out of the business by training schemes and, in some countries, costly provision for early retirement.

A contrary view points out that very much can be achieved by contracting and sharefarming as reviewed in Chapter 4, by collaboration in labour and equipment provision, and cooperation in marketing in which governments can take a facilitating role. There is large scope for the greater application of farmers' skills and knowledge. Two main elements arise frequently in the literature. First, is the need to provide accredited advice on farm business management to smaller farmers, whether individually or in collaborative groups. As the English Policy Commission (2002) remarked: 'Future farmers will sometimes have to be group chief executive, marketing manager, environmentalist and precision grower in the space of a day – or an hour.' This may exaggerate, but successful farmers already have not only to be highly skilled, but also multi-skilled. Second, is the need to reconnect the farmer with the consumer, and vice versa, involving important questions of shortening supply chains, and of rebuilding the image of farming as an honorable and responsible profession, much dented – especially in Europe – by the food-health scares of recent years and by the widely visible consequences of productivist management of the land and its flora and fauna.

Opposition to the productivist system

One imponderable arises from the widespread political opposition to the productivist plans and actions of some agribusiness interests. These have focused in recent years on genetically modified (GM) crops and productivity-enhanced livestock. Seen by their promoters as the way to the future in 'efficient' farming, these technocratic innovations have enjoyed considerable success in North America and Argentina, and also are now well established in some other regions. By contrast, they have met a sustained wall of public opposition in Europe, leading some southern hemisphere and developing countries to oppose adoption for fear of being excluded from European markets.

Consumer resistance has underlain the opposition to GM products, and it has spread internationally even back to North America's domestic markets. Whether or not there is any scientific justification for the fears of food contamination – and no proof has yet emerged – the fears show no sign of going away. There is a more scientifically warranted concern over accidental contamination of non-GM crops by drift of GM pollen, a concern which has increased as organic foods have become more popular and as 'green' politics has acquired significant voter support. The activist destruction of GM trial plots in Britain and France was punished by the courts, but had quite wide public approval, as did the disruption by international activist groups of successive WTO meetings. This contributed to the 2006 deadlock in the Doha round of trade negotiations, and to a growing and widening consumer resistance to the continuing commercial and technological drive in food production and in agriculture as a whole.

The issues coming together politically

We have seen in many places in this book that agriculture generally, and smaller-scale agriculture in particular, everywhere fell on hard times by the 1990s, and in later chapters that these hard times have not gone away. Farmers have lost bargaining power in developed countries as economic weight has concentrated massively both upstream of the farmer and downstream toward the consumer. The most powerful agents have become the seed and chemical multinationals on the one hand, and the big private dealers and major supermarket chains on the other. The banks are no longer as important as they were. It suits these agents (or 'capitals' in another literature) if the farming sector between them remains financially healthy, and hence their political influence has been exercised in support of protective tariffs and subsidies. On the other hand, it would not suit them for farmers to become more powerful as trade partners. They look askance at efforts to form powerful cooperatives as well as at surviving state intervention in trading arrangements which seem to support farmers through such devices as 'single-desk marketing'.[2]

Mainstream farmers' organizations, dominated by larger farmers, have tended to work quite well with the upstream and downstream trading partners of agriculture, and especially with government. Through a large 'black box' of many deals and agreements, they have secured benefits for their (larger) members while fairly consistently supporting productivist policies which have favoured the interests of the upstream and downstream sectors. By the 1980s there was growing dissatisfaction among smaller farmers in Europe and North America over the disproportionate share of subsidy benefits going to the larger farmers and landowners, and over the growing difficulties experienced by smaller farmers, especially those who were indebted. It arose along with the growing dissatisfaction in the developing countries over the sale of subsidized grain and other developed country produce in their markets, at below real cost of production – that is, 'dumped'.

With growing alarm at the activities of the multinational seed and chemical companies,[3] and discontent with the ways of the mainstream organizations, these concerns led to the formation of rival farmers' organizations, of which the best known is the French *Confédération Paysanne* (*La conf*) formed in 1987, one year after the United States National Family Farm Coalition (NFFC). The latter has remained mainly a lobbying group, but *La conf* has taken a more activist route. *La conf* spearheaded the formation of a European Farmers' Coordination body (*Coordination Paysanne Européenne*, CPE), which has followed the American road in acting as a lobbying organization. Together with more than 130 national or provincial organizations of very varying strength and political hue in other countries, both are now members of the international *Via Campesina* movement. Founded in 1992, *Via Campesina* was for some time in effect led by José Bové, a founder of *La Conf* and the CPE, and has consistently adopted policies against neo-liberalism and all its works, against also the WTO, the seed-chemical multinationals and the big food and fibre trading companies. It is now coordinated from Jakarta. Among its members is the Brazilian MST (Chapter 11).

It is a commentary on the polarizing power of neo-liberalism that small family farmers, who nowadays tend to hold politically conservative views even though supporting central marketing and price regulation, should have come to be represented by bodies so strongly on the political left. Had they emerged a few years earlier, they would probably have been branded as communist by their foes, and the activists among them are today sometimes dubbed 'terrorists'. Although with different degrees of force, the member groups of *Via Campesina* all seek better treatment for small farmers in government policies, both financial and legislative. They are against subsidies but demand support prices that will adequately compensate costs of production. They support environmentally sensitive management and the programmes which help fund it, oppose genetic modification of crops and the feeding of livestock with production-enhancing supplements, and oppose free-trade agreements which advantage agribusiness and the multinationals. Some, even the United States NFFC, seek legal prohibition of mergers in agri-business, transportation, food processing and retailing.

The central principle espoused by *Via Campesina* and its members is 'national food sovereignty', protecting the home market for farmers and outlawing all shades of 'dumping', and ensuring fair wages and workers' rights in all agriculture. Food sovereignty includes the right of all nations to develop their own farm and food policies in the interests of their own citizens. While not opposing the need for regulation, therefore, they are against most of the practices of the WTO, and most of them rejoiced at the 2006 impasse in the Doha round of WTO renegotiations. Globalization to the advantage of big capitalism is anathema to them.

Almost none of the participating organizations is a principal leader of opinion or policy formation in its own country, so that the confusing collective programme of *Via Campesina* represents only an opposition. It does nonetheless indicate how the countervailing forces to neo-liberalism have evolved. Except in the guise of redistribution of (subsidy) incomes from the larger farmers to the small, socialism has almost no place, and modern environmentalism, though firmly embraced, appears secondary.

The main thrust is one of reaction, back to protectionism, national self-sufficiency policies, defence against globalization and opposition to multinational agribusiness. Not least is a desire to overthrow the dominance over national policies achieved by the international financial institutions, the WTO itself and the developed country political consensus achieved through such gatherings of national leaders as the G8. In joining the modern reaction against subsidies, the small farmers' organizations are protesting against a system from which their members have received more damage than benefits. In seeking fair prices for their produce – 'fair trade' that is – they are harking back to policies of the period immediately after World War II. Rather unrealistically, they want to achieve this through flexible price management rather than by overt subsidy or any sort of direct payment, seeking essentially the same mechanisms that led ultimately to the huge surpluses of the later twentieth century (Chapter 6). In addition, and in this they do not differ from mainstream farmers' organizations in the North Atlantic countries, they want to be paid 'fair prices' for their whole range of commodities. Payment for environmental services is regarded as necessary and important, but additional.

For developing country farmers, protection from dumped food imports means tariff protection, urged by many observers, but most unlikely to be achieved so long as their governments rely on aid from the developed countries. For farmers in developed countries, who seek limitations to the power of their upstream and downstream partners, the strongest hope is to sustain a loose alliance with environmental and consumer groups which target much the same group of corporations. What developed and developing country movements have in common is a wish to curb the power of the multinational businesses which have gained very substantially from the 1995 agreements that led to the formation of the WTO, and which had hoped for further gain in the contested Doha round. As Harriet Friedmann (1993) pointed out in her insightful paper on the global food regime, the destruction of the Mexican national food organization, CONASUPO, in preparation for the free trade agreement with the US and Canada, offers a rather clear illustration of what the multinationals hope to achieve (Chapter 11).

Friedmann's 1993 paper was written in the context of a then-likely failure of the talks which led to the 1995 world trade agreement. Time has moved on, but much of her argument remains very relevant today. She concluded her paper with a vision of a 'democratic' food regime, in which the people who are producers and consumers, rather than the businesses between and around them, would be paramount. Unlike a system which empowers multinational business, and which emphasizes distance and durability, such a system would emphasize proximity and seasonality: 'what is increasingly clear is that healthy food and environmentally sound agriculture must be rooted in local economies' (Friedmann 1993: 55). In their vision of a revitalized French countryside, Bové and Dufour (2005) said much the same. Rather less positively, and from a mainstream standpoint, so did even the English Policy Commission (2002).

Farmers at work to change the system

In Chapter 13, we introduced the way in which members of the widely distributed Wageningen school have been studying what European farmers, and some consumers, are themselves doing in the same broad direction. The commonest strategy for both surviving and expanding farm household businesses has certainly been a continuation of growing reliance on off-farm and non-farm sources of income and capital first analysed more than 20 years ago by Marsden *et al.* (1986). Years later, van der Ploeg (2000) drew attention to more positive on-farm adaptations arising from the need to reduce farm costs. A reduction in the use of purchased inputs in favour of on-farm recycling, extension of the life of farm equipment and an incremental approach to making changes (thus avoiding the need for substantial credit) are common strategies. Large numbers of conventional farmers, who have no intention of going organic or developing specialized regional production lines, follow these strategies. In a detailed comparison of such 'economical farmers' with conventional, small-intensive and large farmers among Friesian dairy farmers in the 1990s, van der Ploeg found economical farming to be more labour intensive, but also to operate at higher levels of technical efficiency. With this efficiency, use of on-farm inputs and family labour enabled them to control both fixed and variable costs through a period of economic stress arising from loss of the meat market during the BSE crisis, and falling milk prices. They emerged into the twenty-first century with higher surpluses per farm unit than the large farmers, thus putting them in a better position to seize new opportunities, as from the agri-environmental programmes. They had also reduced nitrogen losses to more manageable levels. In the 1960–90 period the low external input strategy had already enabled such farmers to remain viable without entering into the treadmill logic of modernization, and it is therefore unsurprising that economical farming was by 2000 rapidly becoming the dominant style in the Dutch dairy industry.

Van der Ploeg's quantification is rare in the literature, but reduction in the use of external inputs is attractive to many farmers. In Europe and other developed countries there has been extensive adoption of reduced fertilizer and pesticide inputs, though not often going to the extent of evolving a recognizably 'integrated farming system' (Morris and Winter 1999), as mentioned in Chapter 13. It was only full organic farming that attracted state support, and this disciplined alternative appealed only to a minority. Reduction of input costs has become increasingly desirable, even necessary, as prices have risen, but there has been no uniform pattern. Rather, there has been widening diversity of farming styles, one consequence of which has been a very wide range in economic efficiency, measured by costs per unit of production (Policy Commission 2002). To the diversified and pluriactive smaller farmers, production efficiency is not the only goal. The size of the margin between income and outgoings is not necessarily best measured in this way.[4]

In the developing countries, also, there has been no single pattern. Withdrawal of subsidies for fertilizers and pesticides formed a part of the neo-liberal structural adjustment plans, and farmers who had grown accustomed to their generally excessive use then had to 'farm economically' from necessity as prices soared and the official distribution systems were dismantled. The entry of middlemen often made it impossible for farmers to get supplies except through a black market. In northern Nigeria, for example, commercial family farmers experimented by composting town refuse and livestock manure in mixtures with the scarce inorganic fertilizer, thus blending old practices with new (Alexander 1996). Learning by experimenting on soils which they reclaimed from old tin mine dumps, these Hausa farmers learned a great deal about what mixtures to use on different sites, to the extent of recognizing – correctly, as it turned out – that the content of the 15:15:15 Nitrogen: Phosphorus: Potassium (NPK) sold to them by the shops differed significantly between national manufacturers (Pasquini and Alexander 2005). By experimenting, and adapting to their own circumstances, and doing this year by year since the 1970s, they had created by 2001 'almost as many soil fertility management strategies as there are farmers' (Pasquini and Alexander 2005: 115).

In the principal Green Revolution regions of southern and eastern Asia, modern changes have until lately taken place under direction, with provision of seed and 'packages' of fertilizers and pesticides, all tied to credit. In some countries, of which post-1965 Indonesia is a striking example, direction of farmers in the most productive areas was very close. Diversification of crops and farmers' activities has reappeared widely since the mid-1980s as an unintended – but not undesirable – consequence of partial withdrawal of government intervention under neo-liberal dictates and financial austerity. Not only does diversification broaden the income base for the farmers, but there are also social advantages. Many of the vegetable crops that have taken the place of monoculture wheat and rice have a large labour requirement, thus sustaining farm employment even while non-farm jobs were also appearing elsewhere in the region. One ecological consequence of rice monoculture under the Green Revolution has also had an important social spin-off. Massive pest invasions were for a long time combated by the breeding of resistant lines and the heavy use of pesticides. Both strategies proved less and less effective. By degrees, this led to the adaptation of integrated pest management (IPM), encouraging predators of the damaging and disease-bearing insects by creating environment for their habitat, and confining use of pesticides to responsive rather than hopefully preventive use. With FAO support, IPM was introduced in Indonesia in 1986 and at the same time most broad spectrum insecticides were banned from use on rice. By the mid-1990s, there had been a major reduction in insecticide use, but production had continued to grow (Brookfield 2001). As in the Nigerian case just discussed, an important expansion of farmers' knowledge and skills was involved. In addition, the introduction of IPM called on farmers to make decisions, rather than simply following a set of instructions. It had the effect of rebuilding farmers' confidence in their own abilities (Winarto 1995, 2004).

Re-linking the farmer and the consumer

The modern trends in marketing reviewed in Chapter 5 have made the farmer invisible to the consumer, and in truth not many consumers care about where their food comes from, or how it is grown. The farmer's own field of awareness needs go no further than the wholesale trader or the agent who buys for the supermarket. Not only academic and activist writers, but also 'experts' reporting for governments, have seen the reconnection of producer and consumer as a matter of high priority. The mushrooming growth of farmers' markets since the end of the 1980s demonstrated that a minority of farmers and customers themselves also aim for this. These institutions, more purposive than the fresh produce market stalls that were still to be found in European towns as late as 1980, usually permit no middlemen to operate, and often farmers have to be drawn from within a stated distance. Either a family member or an employee must be at the stall to answer customers' questions about the produce on offer. While good personal relations may develop between customers and shopkeepers in those countries and fields of commerce in which family shopkeeping still flourishes, only the farmers' market makes possible direct face-to-face contact between customer and primary producer, or his or her close associate. A relationship of mutual regard may, over time, develop (Kirwan 2004). Together with other means of restoring or creating this relationship, such as farm shops, box systems and – principally in North America – community-supported agriculture (DeLind 2003), these innovations embed the farmer–consumer relationship in a locally or regionally defined society. All of them remain peripheral to the well-organized supply chains of the supermarkets, and cannot challenge the convenience of supermarkets for a majority of customers.

Nonetheless these forms of direct sale are of value to a sufficient number of consumers to make them a viable alternative for the minority of farmers willing and able to diversify their production and practices into lines amenable to direct sale. They thus avoid loss of money to the middleman, and assure themselves of a good retail price for their goods. Other developments are now imminent. The possibilities of electronic sale and advertising can be expected to be used more widely in the near future. Most developed country farmers own computers and even in the developing countries a growing number of farmers have access to Internet terminals. Farmers can both advertise to a targeted audience and arrange direct sales in this way (Box 13.1). Already many farmers in developed countries get marketing information and deal in futures markets on the Internet, and this innovation is potentially equally available to developing country farmers. The entry costs to electronic trading are quite small, so that soon this means of doing business will become as accessible to small farmers as it is to larger farmers, and in most parts of the world.

In recent years, economical farming and direct sale have been strategies of particular benefit to the small-scale farmer, and together they have been responsible for the survival of many family farmers in the developed countries through a period of considerable difficulty. In France, for example, *La conf* candidates for election to the regional chambers of agriculture in 2007 included many small farmers who had diversified and gone in for direct sale.[5] One 55-year-old candidate remarked: 'j'ai pu rester paysan et réaliser mon rêve en réorientant ma production vers la vente directe de legumes et volailles' (I have been able to remain a farmer and realize my dream by reorienting my production to direct sale of vegetables and poultry). Another candidate, a 44-year-old woman with significantly more land, took a wider view in urging that 'parier individuellement et surtout collectivement sur la vente de proximité va devenir de plus en plus néccesaire avec l'augmentation du prix de l'énergie. Dans ce domaine nous avons une carte à jouer avec les consomateurs et les citoyens' (Both individually and collectively, it is important to engage in local direct sale, and this is going to become more and more necessary due to the rising cost of energy. In this we have a card to play with consumers and the citizens) (*Solidarité Paysanne* 2007). The question of 'food miles', the distances between farmers and the ultimate consumers, recently has become a matter of significant political concern in Europe and some other regions. Highly efficient food chains transport produce over long distances, even between continents and across hemispheres. They have facility to search for the lowest-cost producers but there is great expenditure of energy in this mode of transacting business. With the use of energy becoming of great political concern in the context of global warming, this is not an issue likely to die away. Paradoxically, the small farmers' associations that promote such issues are not likely soon to gain louder voice. In France, where *La conf* made substantial gains at the expense of the mainstream body in 2001, they lost these gains in the 2007 elections.

Why family farming persists

In so far as they were projecting forward, rather then interpreting the past, it would seem that Marx, Lenin and even Kautsky were wrong, but nor can it be claimed that Chayanov was altogether right. Modern rural communities are not yet surrounded by a sea of agribusiness mega-farms producing food by remote control (plus the help of gangs of mainly immigrant workers). Nor are they surrounded only by family farms which produce in order to satisfy their own modest needs and retain full decision-making power over what happens within their own gates. Both tendencies seemed strong in the twentieth century, the latter through the land reform movements that persisted until the 1970s and revived in the 1990s, but neither has achieved completion. Nor is this likely in the coming generation, and this is especially true of any projection of the Chayanov model.

The family farms that have survived through the twentieth century have not done so unchanged. Many have grown larger by buying land or renting it in, and thus have been able to utilize the economies of scale that mechanization has made possible. Others, whether or not also enjoying economies of scale, have diversified their production and have made good use of economies of scope by employing the same set of resources to support different activities. Entrepreneurial farm contractors, working on both large and small farms, have created an element that both enlarges and reduces family operation. A great many farm households, large and small, have become pluriactive, so that in many countries, by no means only in the developed lands, the larger part of household incomes is now derived off-farm, yet the farm remains a central household activity. Although there has been substantial re-emergence of self-provisioning on developing country farms, under pressure from rising costs and inadequate or falling prices, only a tiny fraction of farms anywhere now produces only for household subsistence. Almost all produce at least some commodities for the market. What Marxisant writers call 'simple commodity production' or 'petty commodity production' characterizes most family farms in all parts of the world. And there is nothing simple about it, nor in sum is it 'petty'.

Yet a great number of farms have gone, and their number continues to diminish. However optimistic we may be about the future of family farmers, we have to recognize that by 2034 there will be fewer of them than there are today. The movement toward shorter supply chains and local marketing, and the much more problematic renewed emphasis on national food security, will feel the force of the growing number of inter-country and regional free trade agreements. Trading associations between developed and developing countries are resulting in reduction of tariff and other forms of protection. There is no reversal of the globalizing trend described, along with others, by Bernstein (2006: 8) who writes that 'agriculture is increasingly, if unevenly, integrated, organized and regulated by the relations between agrarian classes and types of farms, on the one hand, and (often highly concentrated) capital upstream and downstream of farming, on the other hand'.

Bernstein is following up his earlier suggestion, mentioned in Chapter 11, that an 'agrarian question of labour' has, since the 1970s, taken the place of the old 'agrarian question of capital' (Bernstein 2004). He is arguing that globalization has accelerated fragmentation of the labour force, especially but not only in the developing countries, so that 'most have to pursue their means of livelihood/ reproduction across different sites of the social division of labour: urban and rural, agricultural and non-agricultural, wage-employment and self-employment' (Bernstein 2006: 9). In this still-Marxisant view, all this is enforced on society as part of the operation of late capitalism. It involves an involuntary blurring of the old division of labour, including that which was celebrated famously in the revolutionary alliance between the hammer and the sickle – workers and peasants. As in the presentation of de-agrarianization (Chapter 12), we find reason to question the top-down political economy view of a process which, in any case, is very variably expressed on the landscape. In regard to de-agrarianization, opportunity had a major place in its explanation, and we believe that the same applies to the 'fragmentation' of the labour market. The fundamental change in mobility of individuals produced by the internal combustion engine has widened choices for rural workers every-where during the last hundred years. The fact that images of other places and peoples, however selective, can now appear on television screens in the most remote villages and farms in all lands has widened that comparison of perceived living standards which Collantes (2004) regarded as fundamental to an explanation of rural depopulation. It is also fundamental to the widening of information fields that underlies the break-up of the categories of 'worker and peasant' and 'rural and urban'. Not least, it is potentially damaging to the integrity of the family household and its farm.

What sort of farm a generation hence?

Bernstein does have his finger on one important issue. Although land is obviously the basis of all farming, the capitalist economy no longer requires the separation of labour from its land in order to control that labour and its product. The markets for produce and labour, and even services, are now so comprehensively developed that effective control is possible even if the worker fully retains his land. Control may be exercised by private concerns upstream and especially downstream of the farm, and by government. Bernstein focuses his attention on labour; others have focused on control of marketing (as in contract farming), input supply (as in the works of the seed-chemical industry) or the invasiveness now possible for governments in all countries with elaborated systems of telecommunications and surveillance of citizen behaviour.

Fragmentation of the labour market, as described by Bernstein, is the outstanding expression of the individualization tendencies inherent in capitalism. It could break up the farm family household in adult life, and the very numerous reports of unwillingness of children to inherit the farm, except as property for sale, suggest that this is precisely what may be happening. Pluriactivity can undo many of the practical advantages of extended- and even nuclear-household farm operation. The successfully adapting and competing Chayanovian household farm can become simply the common residence of a small number of people working in different trades and different places, only one of whom is the farmer. This internal disintegration has happened. On the other hand, a substantial proportion of large farms in one European region, southeastern England, were divided through multiple succession over the past 40 years (Burton and Walford 2005). After division, they grew again by purchase and renting of land, but the farms were renewed. Taking account also of Roberts's (1996) discussion of the reproduction of family farms in the great plains of the USA, there are grounds for believing that farm renewal may be more widespread than is often thought.

Resolving contradictions

In this book, and especially in Chapter 13 and in this chapter, we have sought to emphasize the ways in which the landholding farmer endeavours to resist external control. All involve ways of revalorizing farming as a profession, and all are structured around the family farm and its household. Efforts to farm economically, which underlie most of them, require use of family labour much on the lines theorized by Chayanov and Schmitt (1991). There is thus a conflict, or 'contradiction' in Marxist terminology, between strategies seeking to revitalize family farming and individualizing tendencies that seem to deny the possibilities of success.

This issue, more than that of subsumption of the family farm by direct competition from capitalist producers, or indirect subsumption by the capitalist input producers and controllers of the output supply chains, seems to us to be a central question for the coming quarter-century. With machinery now manufactured and sold in a wide range of sizes, the scale issue no more rules against family-level operation than it did 130 years ago, when family farms successfully competed against capitalists in commercial cereal production (Friedmann 1978a, 1978b). Use of economical farming with its technical efficiencies restores some competitive advantages to the smaller farmers so that economic efficiency can become, to a great degree and over a wide range of farm size, independent of the scale of operation.

Family farming can be much more flexible in its responses to both constraints and opportunities than company-managed farming with its heavy capitalization and paid workforces. The substantial shift from regular to gang labour among the latter is, in large measure, a response to a need to regain flexibility in the face of conditions which make the going tough. But labour gangs are a blunt instrument by comparison with the more finely honed responses that family-centred farms can achieve. Often in the rapidly changing economic environment of the twentieth century, it has been the small farmers who have been the entrepreneurs in innovating with new crops and new practices, paving the way for the larger farmers, and their labour gangs, to follow. One central question for family farms thus becomes their ability to survive forces promoting internal fragmentation.

The issue therefore moves from the material field, the central province of economics, to the social. The individualizing tendencies of the market economy as a whole here run into an obstacle. Everywhere, and at all times, the economy is embedded within social institutions. Among these, the most widespread and enduring has been the family in both its nuclear and extended forms. Neither doctrinaire attempts to break this down in collectivizing nations nor the informal-ization of sexual relations in much of the modern world have done more than dent institutions in which the cement has been the enduring interpersonal relationships of parenthood, siblings and close affines. Certainly the family relationship can be broken, but the expectation that it can endure through time is that which underlies the concept of the family farm. What Gray (1998) seems to have meant when he wrote of the consubstantiation of the family and its farm is that a kinship group of size to manage a farm of a certain size expects to go on doing so for a long period, enlarging or reducing it as necessary, changing its enterprises as necessary or desirable. Even as economic forces tend toward fragmentation, the social unit has, in the main, survived. Farms, too, are renewed as we saw above.

A mixed, but not pessimistic conclusion

In a few countries, especially in Africa, the number of farmers continues to increase, but in most regions they are in decline and it seems certain that in 27 years' time there will already be significantly fewer farmers in the world than there are today. There will be a further decline in the following generation. The almost universal spread of pluriactivity not only means that there will also be fewer full-time farmers, but that most certainly there will be many fewer remaining full-time farming house-holds, in developing or developed countries. Given progress toward achievement of Smil's (2001) improvements in technical efficiency, this need not cause concern over food production. But to those who seek sustainable farming, fresh food made available along short supply chains and an agriculture which manages the environment and keeps it in good condition, the long-term trends in farm demography are of serious concern.

On the dark side there is a cluster of spreading problems. Where are the new farmers to come from, especially in the developed countries? The privileged recruits, sons and daughters of farmers, are clearly inadequate in number to fill all the vacancies, and many opt for work other than farming. There is also the gender factor. The female half of this population is, in most societies, still marginalized close to exclusion by the ethos of masculine farm control, an ethos which mechanization has even enhanced (e.g. Saugeres 2002). Female-headed farms are now common in the developing countries, but there are not many in the developed lands. Contractors are certainly a source, as many end up with their own farms. But the population of skilled and experienced farm labourers, who in the past might have gone up the traditional 'ladder' through sharecropping to tenancy, has been almost eliminated by redundancies. There is now no strong political drive to provide landless labourers with farms through land reform. Brazil, where they coexist with large and underused landholdings, is almost unique in having seen them become a major new source of family farmers in the last 25 years. Other constraints are the high costs of entry, as land has increased in value even where there are no subsidies to inflate its price. Except through agricultural colleges, which nowadays recruit diminishing numbers, and some apprenticeship schemes, there are few ways in which the skills and knowledge so essential for good farming can readily be acquired. While farming can be learned, the best time to learn it is while young, and the widespread unpopularity of farm work as a career option means that young recruits from outside agriculture are scarce. Perhaps scarcity will improve remuneration for learner-workers on larger family farms, but not while gangs remain available.

There are, nonetheless, other trends. In many countries, owning or renting a farm offers security and a measure of social prestige. Immigrants, once they can establish themselves, continue to take up small farms, largely for fruit and vegetable production. Not only are such newcomers willing learners, but many already have farming skills which they can adapt to different conditions and lines of production that are wholly new to them. There is also no indication that the supply of ex-urban part-time farmers is drying up and some of these have significant capital to invest. Although the productivity of newcomers from outside agriculture is individually small, collectively it is increasingly significant. Some of them are successful in farming, just as a proportion of established family farmers manage to contain costs, diversify, seize opportunities and achieve good incomes.

The world cannot afford to lose the skills and abilities of its family farmers, and nor can many individual countries, notwithstanding the views of the shallowly economistic writers referred to above. So long as a sufficient number of farmers can sustain their living standards, by means among which pluriactivity is now the most important, they will remain on the land and vacancies will be filled. New farm families still take the place of many of those who depart, and farms continue to grow larger gradually, not through sudden takeover by industrial organizations. Although family farms do continue to be absorbed into large neighbouring capitalist holdings, especially in the Americas, world-wide they are perhaps less under threat from the primitive accumulation of capital today than at any time in the last three centuries. Agriculture is not under most circumstances a promising field for capitalist investment. Risks are too high, and returns too low. The flexibility of family farm organization can cope better with adverse conditions, and enables these smaller units to take advantage of new opportunities.

We can now improve on the gloomy interim conclusion which we reached at the end of Chapter 8. Despite the structural problems which they face, family farms can better cope with a loss of production subsidies than the larger farms. This has been incontrovertibly demonstrated, because most smaller farms have received little benefit from the subsidies of the last half-century and yet so many have survived. We expect direct and production subsidies to be reduced in the life of the coming generation, but not to be extinguished so long as the American farm lobby can retain its influence over US policy.[6] Commercial family farms in other lands have survived without direct subsidies, and in the New Zealand case through the actual and sudden loss of subsidies. Farming without subsidies is quite a realistic proposition.

The view that large scale is critically important, sustained from Marx through modern neo-classical economics, blinds decision makers to the adaptable flexibility of family-scale operation, its competitive ability and to its contribution to sustaining wider rural economies. If production subsidies come to an end and are replaced only by support for the environmental services farmers provide, this will not be to the advantage of the large-scale capitalist farms. It may change public outlook. It may lead to renewed appreciation of the ancient social institution of the family farm, perhaps the oldest of all productive institutions. Nonetheless, it is idle to pretend that visions such as that of Friedmann (1993) and of *Via Campesina* and its member organizations are likely to achieve commanding political weight within the span of the coming generation. The world of our vision, too, is different from that of the economists and others who advise governments and the WTO; it is founded on respect for the needs of the individual producer and consumer and of the environment. But it faces a huge mass of international public indifference.

It is this indifference that needs to be tackled. It has grown substantially through the years of frequent food scares and evident environmental degradation. Many family farmers, but few large-scale capitalist farmers, have responded to the need to change agricultural and pastoral practices. The political representatives of the smaller farmers, in both developed and developing countries, need to stress the way in which family farmers are adapting to the newer public needs, rather than demanding the impossible in terms of flexible price regulation and national food autarky. We can hope that the counter-movement to the starkness of neo-liberalism will gain strength in the coming years, but it will certainly take all the years of the coming generation to achieve this. As the struggle goes on, those who have flexibility to adapt are the most resilient. Family farmers have endured for centuries through their flexibility and resilience. In the uncertainties of the twenty-first century, this resilience remains their strength. Writers in the 2030s can tell their readers how wrong, or right, we have been.

Notes

1   We do not doubt that much higher crop yields are attainable by means of genetic engineering, and note that important work on genetic modification of rice, aimed at changing the photosynthesis path, is being done by the International Rice Research Institute. But most genetic modification undertaken to date has had more commercial ends, and the claims made by the corporations that have undertaken this work are not yet substantiated by results, notwithstanding their appeal to a large number of politicians.
2   Particularly targeted are the 'single-desk' arrangements for wheat in Canada and Australia, both popular with smaller farmers because they offer them prices equal to those obtained by the larger growers. In both countries, these survivors from an earlier period of state involvement in marketing came under political attack in 2007. Neither was likely to survive long.
3   The strongest opposition has been to the 'terminator' technology, designed to make seeds unviable without the application of proprietory chemical treatment, and thus forcing farmers to buy new seed every season and abandon the common and wide-spread practice of saving seed from the previous harvest. Opposition has ensured that this has not yet been employed, but the technology remains and it has obvious attractions to the seed and chemical industry. The need to buy new hybrid seed each season has been accepted for more than half a century. This attempt to force all farmers to buy seed for each crop, and across a range of crops for which saved seed, or seed purchased from or exchanged with neighbours, is viable and an ancient means of saving costs, has generated enormous hostility around the issue of farmers' rights. The topic is discussed more fully in Brookfield (2001: 262–5).
4   Schmitt (1991) usefully points out that a pluriactive farm family can allocate its time between on-farm and off-farm activities in a highly efficient manner. This advantage is not available to a wage-labour farm.
5   The regional chambers of agriculture in France exert considerable power, especially in regulating the land market and administering the CAP and national policies at local level.
6   This influence rests heavily on the close balance between the two main political parties in the USA, leading to quite frequent shifts in many states from one party to the other, following the transfer of only quite a small proportion of votes.

References

Adams, A. (1986) An open letter to a young researcher, in R. Apthorpe and A. Krahl (eds) *Development Studies: Critique and Renewal*, Leiden: E.J. Brill, pp. 218–49.

Adams, J. (1988) The decoupling of farm and household: differential consequences of capitalist development on southern Illinois and Third World family farms. *Comparative Studies in Society and History* 30: 453–82.

Adams, J. (2003) Introduction, in J. Adams (ed.) *Fighting for the Farm: Rural America Transformed*, Philadelphia: University of Pennsylvania Press, pp. 1–21.

Adams, J., Kraft, S., Ruhl, J.B., Lant, C., Loftus, T. and Duran, L. (2005) Watershed planning: pseudo-democracy and its alternatives – the case of the Cache river wetland, Illinois. *Agriculture and Human Values* 22: 327–38.

Agrawal, A. (1998) *Greener Pastures: Politics, Markets and Community among a Migrant Pastoral People*, Durham, NC: Duke University Press.

Albion, R.G. (1939) *The Rise of New York Port [1815–1860]*, New York: Charles Scribner's Sons.

Alexander, J. (1987) *Trade, Traders and Trading in Rural Java*, Singapore: Oxford University Press.

Alexander J. and Booth, A. (1992) The service sector, in A. Booth (ed.) *The Oil Boom and After: Indonesian Economic Policy and Performance in the Soeharto Era*, Singapore: Oxford University Press, pp. 283–319.

Alexander, M.J. (1996) The effectiveness of small-scale irrigated agriculture in the reclamation of mine-land soils on the Jos plateau of Nigeria. *Land Degradation and Development* 7(1): 77–84.

Allan, W. (1965) *The African Husbandman*, Edinburgh: Oliver and Boyd.

Allen, A.W. and Vandever, M.W. (2003) *A National Survey of Conservation Reserve Program (CRP) Participants on Environmental Effects, Wildlife Issues, and Vegetation Management on Program Lands*, US Geological Survey, Biological Science Report, Fort Collins, CO: USGS.

Allen, P. (1999) Reweaving the food security safety net: mediating entitlement and entrepreneurship. *Agriculture and Human Values* 16(2): 117–29.

Almekinders, C. and Louwaars, N. (1999) *Farmers' Seed Production: New Approaches and Practices*, London: Intermediate Technology Publications.

Altieri, M.A. (1995) *Agroecology: The Science of Sustainable Agriculture*, Boulder, CO: Westview Press.

Amanor, K. (1994) *The New Frontier: Farmer Responses to Land Degradation*, London: Zed Books.

Amanor, K. and Diderutuah, M. (2001) *Share Contracts in the Oil Palm and Citrus Belt of Ghana*, London: International Institute for Environment and Development.

Anderson, J. (2004) Dairy deregulation in northern Queensland: the end of traditional farming? *Anthropological Forum* 14: 269–82.

Anderson, T. and Hill, P. (1975) The evolution of property rights: a study of the American west. *Journal of Law and Economics* 43(1): 163–79.

Anon (2004) Obituary: William Hinton, *The Economist* 8377, 29 May: 85.

Argent, N. (2002) From pillar to post? In search of the post-productivist countryside in Australia. *Australian Geographer* 33: 97–114.

Arriaga Jordán, C., Gonzáles Diaz, J., Gonzáles Esquivel, C., Narva Bernal, G. and Valázquez Beltrán, L. (1997) Caracterización de los sistemas de producción campesinos en dos zonas del municipio de San Felipe del Progreso, México: estrategias contrastantes, in G.R. Herrejón, A.A. Hernández, L. Gonzáles Diaz and C. Arriaga Jordán (eds) *Investigación Para el Desarollo Rural: Diez Años de Experiencia del CICA*, Toluca, México: Centro de Investigación en Ciencias Agropecuarias, Universidad Autónoma del Estado de México, pp. 171–225.

Asano-Tamanoi, M. (1988) Farmers, industries and the state: the culture of contract farming in Spain and Japan. *Comparative Studies in Society and History* 30: 432–52.

Ash, R. (2006) Squeezing the peasants: grain extraction, food consumption and rural living standards in Mao's China. *China Quarterly* 187: 959–98.

Bahiigwa, G., Mdoe, N. and Ellis, F. (2005) Livelihoods research findings and agriculture-led growth. *IDS Bulletin* 36(2): 115–20.

Balfour, E.B. (1943) *The Living Soil: Evidence of the Importance to Human Health of Soil Vitality, with Special Reference to National Planning*, London: Faber & Faber.

Banks, J. and Marsden, T. (2001) The nature of rural development: the organic potential. *Journal of Environmental Policy and Planning* 3(2): 103–21.

Barkin, D. (2002) The reconstruction of a modern Mexican peasantry. *Journal of Peasant Studies* 30(1): 73–90.

Barkin, D. and King, T. (1970) *Regional Economic Development: The River Basin Approach in Mexico*, Cambridge: Cambridge University Press.

Barnett, B.J. (2003) The U.S. farm financial crisis of the 1980s, in J. Adams (ed.) *Fighting for the Farm: Rural America Transformed*, Philadelphia: University of Pennsylvania Press, pp. 160–71.

Barraclough, S.L. and Domike, A.L. (1970) Agrarian structure in seven Latin American countries, in R. Stavenhagen (ed.) *Agrarian Problems and Peasant Movements in Latin America*, New York: Doubleday Anchor, pp. 41–94.

Barrett, C.B. (1997) Food market liberalization and trader entry: evidence from Madagascar. *World Development* 25: 763–77.

Barrett, C.B. and Mutambatsere, E. (2006) Marketing boards, in L.E. Blume and S.N. Durlauf (eds) *The New Palgrave Dictionary of Economics*, second edition, London: Palgrave Macmillan.

Bassett, T. (2001) *The Peasant Cotton Revolution in West Africa: Côte d'Ivoire 1880–1995*, Cambridge: Cambridge University Press.

Batie, S.S. (1985) Soil conservation in the 1980s: a historical perspective. *Agricultural History* 59: 107–23.

Batterbury, S.J. (1998) Local environmental management, land degradation and the 'Gestion de Terroirs' approach in West Africa: policies and pitfalls. *Journal of International Development* 10: 871–98.

Batterbury, S.J. (2005) Within and beyond territories: a comparison of village land use management and livelihood diversification in Burkina Faso and southwest Niger, in Q. Gausset, M.A. Whyte and T. Birch-Thomsen (eds) *Beyond Territory and Scarcity: Exploring Conflicts over Natural Resource Management*, Stockholm: Afrika Institutet, pp. 149–67.

Bayliss-Smith, T.P. (1982) *The Ecology of Agricultural Systems*, Cambridge: Cambridge University Press.

Beaufoy, G., Jennings, S., Hernandez, E., Peiteado, C. and Fuentelsaz, F. (2005) *The Spain National Report to Europe's Living Countryside*, Netherlands: WWF Europe, Stichting Natuur en Milieu; and UK: Land Use Policy Group.

Begat, N. (2001) L'eau sale de Bretagne. Les Pénélopes (Ressources), 1 November. Online. Available at <http://www.penelopes.org/article.php3?id_article=1479> (accessed 14 Feb. 2006).

Beinart, W. (1994) *Twentieth-century South Africa*, Oxford: Oxford University Press.

Bellisario, A. (2007) The Chilean agrarian transformation: agrarian reform and capitalist 'partial' counter-agrarian reform, 1964–1980. Part I. Reformism, socialism and free-market neoliberalism. *Journal of Agrarian Change* 7(1): 1–34.

Berg, E. (1989) The liberalization of rice marketing in Madagascar. *World Development* 17: 719–28.

Bernstein, H. (2004) 'Changing before our very eyes': agrarian questions and the politics of land in capitalism today. *Journal of Agrarian Change* 4: 190–225.

Bernstein, H. (2006) 'Is there an agrarian question in the twenty-first century?' Keynote address at Canadian Association for the Study of International Development (CASID) conference, Toronto, June.

Berry, R.A. and Cline, W.R. (1979) *Agrarian Structure and Productivity in Developing Countries*, Baltimore, MD and London: Johns Hopkins University Press.

Blackwood, E. (1997) Women, land and labor: negotiating clientage and kinship in a Minangkabau peasant community. *Ethnology* 36(4): 277–93.

Blaikie, P. and Brookfield, H. (1987) *Land Degradation and Society*, London: Methuen.

Blanco, H. (1972) *Land or Death: The Peasant Struggle in Peru*, New York: Pathfinder Press.

Boardman, J. and Poesen, J. (2006) *Soil Erosion in Europe*, Chichester: John Wiley.

Bobrow-Strain, A. (2004) (Dis)accords: the politics of market-assisted land reforms in Chiapas, Mexico. *World Development* 32: 887–903.

Bonnemaison, J. (1967) *Le Terroir se Tsarahohenana: Introduction à la Region d'Ambohibary (Vikinankaratra)*, Tananarive: Centre ORSTOM (Office de la Recherche Scientifique et Technique Outre-Mer).

Borras, S.M. Jr (2005) Can redistributive reform be achieved via market-based voluntary land transfer schemes? Evidence and lessons from the Philippines. *Journal of Development Studies* 41: 90–134.

Bové, J. and Dufour, F. (2005) *Food for the Future: Agriculture for a Global Age*, Cambridge: Polity Press.

Bramall, C. (2004) Chinese land reform in long-run perspective and in the wider east-Asian context. *Journal of Agrarian Change* 4: 107–41.

Bramall, C. (2006) The last of the romantics? Maoist economic development in retrospect. *China Quarterly* 187: 686–92.

Brass, T. (2005) Neoliberalism and the rise of (peasant) nations within the nation: Chiapas in comparative and theoretical perspective. *Journal of Peasant Studies* 32: 651–91.

Brons, J.E. (2005) *Activity Differentiation in Rural Livelihoods: The Role of Farm Supplementary Income in Burkina Faso*, Tropical Resource Management Papers 66, Wageningen, the Netherlands: Wageningen University and Research Centre.

Brookfield, H. (1969) Introduction: the market place, in H. Brookfield (ed.) *Pacific Market Places: A Collection of Essays*, Canberra: Australian National University Press, pp. 1–24.

Brookfield, H. (1972) Intensification and disintensification in Pacific agriculture: a theoretical approach. *Pacific Viewpoint* 13: 30–48.

Brookfield, H. (1975) *Interdependent Development*, London: Methuen.

Brookfield, H. (1994) *Transformation with Industrialization in Peninsular Malaysia*, Kuala Lumpur: Oxford University Press.

Brookfield, H. (2001) *Exploring Agrodiversity*, New York: Columbia University Press.

Brookfield, H. and Brown, P. (1963) *Struggle for Land: Agriculture and Group Territories among the Chimbu of the New Guinea Highlands*, Melbourne: Oxford University Press.

Brookfield, H. with Hart, D. (1971) *Melanesia: A Geographical Interpretation of an Island World*, London: Methuen.

Brookfield, H. and Padoch, C. (2007) Managing biodiversity in spatially and temporally complex agricultural landscapes, in D.I. Jarvis, C. Padoch and H.D. Cooper (eds) *Managing Biodiversity in Agricultural Ecosystems*, New York: Columbia University Press, pp. 338–61.

Brookfield, H., Samad Hadi, A. and Zaharah Mahmud (1991) *The City in the Village: The In-situ Urbanization of Villages, Villagers and their Land around Kuala Lumpur, Malaysia*, Singapore: Oxford University Press.

Brookfield, H., Padoch, C., Parsons, H. and Stocking, M. (eds) (2002) *Cultivating Biodiversity: Understanding, Analysing and Using Agricultural Diversity*, London: ITDG Publishing.

Brookfield, H., Parsons, H. and Brookfield, M. (eds) (2003) *Agrodiversity: Learning from Farmers across the World*, Tokyo: United Nations University Press.

Brown, P. and Brookfield, H. (2005) Sweet potatoes, pigs and the Chimbu, in C. Ballard, P. Brown, R.M. Bourke and T. Harwood (eds) *The Sweet Potato in Oceania: A Reappraisal*, Pittsburgh and Sydney: Ethnology Monographs 19 and Oceania Monographs 56, pp. 131–6.

Brown, P., Brookfield, H. and Grau, R. (1990) Land tenure and transfer in Chimbu, Papua New Guinea: 1958–1984 – a study in continuity and change, accommodation and opportunism. *Human Ecology* 18: 21–49.

Brush, S.B. (1999) The issues of *in situ* conservation of crop genetic resources, in S.B. Brush (ed.) *Genes in the Field: On-farm Conservation of Crop Diversity*, Boca Raton, FL: Lewis Publishers, pp. 3–26.

Brush, S.B., Tadesse, D. and Van Dusen, E. (2003) Crop diversity in peasant and industrialized agriculture: Mexico and California. *Society and Natural Resources* 16: 123–41.

Bryceson, D.F. (1996) Deagrarianization and rural employment in sub-Saharan Africa: a sectoral perspective. *World Development* 24: 97–111.

Bryceson, D.F. (1997) De-agrarianisation in southern Africa: acknowledging the inevitable, in D.F. Bryceson and V. Jamal (eds) *Farewell to Farms: Deagrarianisation and Employment in Africa*, Leiden: African Studies Centre, pp. 237–56 and Research Series 1997/10, Aldershot: Ashgate.

Bryceson, D.F. (2002) The scramble in Africa: reorienting rural livelihoods. *World Development* 30: 725–39.

Buck, D., Getz, C. and Guthman, J. (1997) From farm to table: the organic vegetable commodity chain of northern California. *Sociologia Ruralis* 37(1): 3–20.

Bureau, J-C. (2003) Enlargement and reform of the EU Common Agricultural Policy: impacts on the western hemisphere countries, Washington, DC: Inter-American Development Bank (mimeo).

Burton, R.J.F. (2004) Seeing through the 'good farmer's' eyes: towards developing an understanding of the social symbolic value of 'productivist' behaviour. *Sociologia Ruralis* 44: 195–215.

Burton, R.J.F. and Walford, N. (2005) Multiple succession and land division on family farms in the south east of England: a counterbalance to agricultural concentration? *Journal of Rural Studies* 21(3): 335–47.

Burton, R.J.F. and Wilson, G.A. (2006) Injecting social psychology theory into conceptualizations of agricultural agency: toward a post-productivist farmer self-identity? *Journal of Rural Studies* 22: 95–115.

Buttel, F.H. (2001) Some reflections on late twentieth century agrarian political economy. *Sociologia Ruralis* 41: 165–81.

Byres, T.J. (2004) Neo-classical neo-populism 25 years on: déjà vu and déjà passé. *Journal of Agrarian Change* 4: 17–44.

Campbell, H. and Liepins, R. (2001) Naming organics: understanding organic standards in New Zealand as a discursive field. *Sociologia Ruralis* 41(1): 21–39.

Carney, D. (ed.) (1998) *Sustainable Rural Livelihoods: What Contribution Can We Make?*, London: Department for International Development.

Carson, R. (1962) *Silent Spring*, Boston, MA: Houghton Mifflin.

Castelán Ortega, O., Gonzáles Esquivel, C., Arriaga Jordán, C. and Chávez Meija, C. (2003) Mexico, in H. Brookfield, H. Parsons and M. Brookfield (eds) *Agrodiversity: Learning from Farmers across the World*, Tokyo: United Nations University Press, pp. 249–69.

Centre for Rural Economics Research and CJC Consulting (2002) *Economic Evaluation of Agri-Environmental Schemes*, Cambridge: Department of Land Economy.

Chambers, R. (1994) Participatory rural appraisal (PRA): challenges, potential and paradigm. *World Development* 22: 1437–54.

Chambers, R. (1997) *Whose Reality Counts? Putting the First Last*, London: Intermediate Technology Publications.

Chambers, R., Pacey, A. and Thrupp, L.A. (1989) *Farmer First: Innovation and Agricultural Research*, London: Intermediate Technology Publications.

Chan, A., Madsen, R. and Unger, J. (1984) *Chen Village: The Recent History of a Peasant Community in Mao's China*, Berkeley and Los Angeles: University of California Press.

Chayanov [Tschajanow], A.V. (1923) *Die Lehre der bäuernilichen Wirtschhaft*, Berlin: P. Parey.

Chayanov, A.V. (1925) *Organizatsiya Krest'yanskogo Khozyaistva*, Moskva: Tsentral'noe Tovarichestvo Kooprativnogo.

Chayanov, A.V. (1966) *The Theory of Peasant Economy*, edited by D. Thorner, B. Kerblay and R.E.F. Smith, Homewood, IL: Irwin for the American Economic Association.

Chenery, H., Ahluwalia, M.S., Bell, C.L.G., Duloy, J.H. and Jolly, R. (1974) *Redistribution with Growth*, London: Oxford University Press.

Chevalier, F. (1952) *La Formation des Grandes Domaines au Mexique: Terre et société aux XVIe–XVIIe Siècles*, Paris: Institut d'Ethnologie, trans by A. Eustis (1963) *Land and Society in Colonial Mexico: The Great Hacienda*, Berkeley: University of California Press.

Chi, S.C. and Taylor, M. (1986) Business organisations, labour demand and industrialisation in Peninsular Malaysia, in T.G. McGee (ed.) *Industrialisation and Labour Force Processes: A Case Study of Peninsular Malaysia*, Canberra: Research School of Pacific Studies, Australian National University, pp. 41–77.

Chimhowu, A. and Woodhouse, P. (2006) Customary vs. private property rights? Dynamics and trajectories of vernacular land markets in sub-Saharan Africa. *Journal of Agrarian Change* 6: 346–71.

Christie, J.W. (2004) The agricultural economies of early Java and Bali, in P. Boomgard and D. Henley (eds) *Smallholders and Stockbreeders: History of Foodcrop and Livestock Farming in Southeast Asia*, Leiden: KITLV Press, pp. 47–67.

CIDA (1971) *Tenencia de la Tierra y Reforma Agraria en América Latina: Informe Regional y Resúmenes de los Estudios por Países*, Washington: Comité Interamericano de Desarollo Agrícola.

Classen, R., Hansen, L., Peters, R., Brenneman, M.,Weinberg, M., Cattaneo, A., *et al.* (2001) *Agri-environmental Policy at the Crossroads: Guideposts in a Changing Landscape*, Economic Research Service, AER 794, Washington: US Department of Agriculture.

Cloke, P. (1996) Looking through European eyes? A re-evaluation of agricultural deregulation in New Zealand. *Sociologia Ruralis* 36: 308–30.

Coatsworth, J.H. (2005) Structures, endowments, and institutions in the economic history of Latin America. *Latin American Research Review* 40: 126–44.

Collantes, F. (2004) La evolución de la actividad agrícola en las áreas de montaña españolas (1860–2000). *Revista de Estudios Agrosociales y Pesqueros* 201: 79–104.

Collantes, F. (2005a) Farewell to the peasant republic: the demise of traditional economies in marginal Europe (1850–1990), paper presented at Economic History Society Annual Conference, Leicester, UK, 9–11 April.

Collantes, F. (2005b) Declive demográfico y cambio económico en las áreas de montaña españolas (1860–2000). *Revista de Historia Económica* 23(3): 515–40.

Collantes, F. (2006a) The evolution of rural provisioning institutions in an industrialising economy: peasants and capitalism in twentieth-century Spain, paper presented at Fourteenth International Economic History Congress, Helsinki, Finland, 21–25 August.

Collantes, F. (2006b) Rural occupational change in industrializing Europe: a comparison of England, France and Spain, paper presented at meeting of the Development Economics Study Group (Derg), University of Copenhagen, 13 Sept.

Committee on the Role of Alternative Farming Methods in Modern Production Agriculture (1989) *Alternative Agriculture*, Washington: National Academy Press.

Conchol, J. (1973) The agrarian policy of the popular government, in A. Zammit (ed.) *The Chilean Road to Socialism*, Brighton: Institute of Development Studies at the University of Sussex, pp. 107–14.

Constance, D.H., Kleiner, A.M. and Rikoon, J.S. (2003) The contested terrain of swine production: deregulation and reregulation of corporate farming laws in Missouri, in J. Adams (ed.) *Fighting for the Farm: Rural America Transformed*, Philadelphia: University of Pennsylvania Press, pp. 75–95.

Cook, R. (2005) Supermarket challenges and opportunities for producers and shippers: US experience. *Farm Policy Journal* 2(1): 46–52.

Coombs, B. and Campbell, H. (1998) Dependent reproduction of alternative modes of agriculture: organic farming in New Zealand. *Sociologia Ruralis* 38(2): 127–45.

Corner, L. (1983) The persistence of poverty: rural development policy in Malaysia. *Kajian Malaysia* 1: 38–59.

Crabtree, J. (2002) The impact of neo-liberal economics on Peruvian peasant agriculture in the 1990s. *Journal of Peasant Studies* 29(3–4): 131–61.

Craig, D. and Porter, D. (2006) *Development Beyond Neoliberalism? Governance, Poverty Reduction and Political Economy*, London: Routledge.

Cullen, P., Williams, J. and Curtis, A. (2003) *Landcare Farming: Securing the Future for Australian Agriculture*, Chatswood, NSW: Landcare Australia.

Cumberland, C.C. (1968) *Mexico: The Struggle for Modernity*, New York: Oxford University Press.

Cunningham, I. (2005) *The Land of Flowers: An Australian Environment on the Brink*, Brighton Le Sands, NSW, Australia: Oxford Press.

Daley, E. (2005) Land and social change in a Tanzanian village 1: Kinyanambo 1920s–1990; 2: Kinyanambo in the 1990s. *Journal of Agrarian Change* 5: 363–404 and 526–72.

Dao, Z., Guo, H., Chen, A. and Fu, Y. (2003) China, in H. Brookfield, H. Parsons and M. Brookfield (eds) *Agrodiversity: Learning from Farmers across the World*, Tokyo: United Nations University Press, pp. 195–211.

Darnhofer, I. (2005) Organic farming and rural development: some evidence from Austria. *Sociologia Ruralis* 45: 308–23.

Darnhofer, I., Schneeberger, W. and Freyer, B. (2005) Converting or not converting to organic farming in Austria: farmer types and their rationale. *Agriculture and Human Values* 22: 39–52.

Davis, J.H. and Goldberg, R.A. (1957) *A Concept of Agribusiness*, Boston, MA: Alpine Press for Harvard University.

de Anda, G. (1974) *El Cardenismo: Desviación Totalitaria de la Revolución Mexicana*, Mexico, DF: privately published.

de Roest, K. and Menghi, A. (2000) Reconsidering 'traditional' food: the case of Parmigiano Reggiano cheese. *Sociologia Rurales* 40(4): 439–51.

Debailleul, G. (2002) Eléments d'analyse comparative des réglementations environnementales en matière d'élevage intensif. Document de référence, Québec: Université Laval.

Deininger, K. (2003) *Land Policies for Growth and Poverty Reduction*, A World Bank Policy Research Report, New York and Oxford: Oxford University Press.

DeLind, L.B. (2003) Considerably more than vegetables, a lot less than community: the dilemma of community-supported agriculture, in J. Adams (ed.) *Fighting for the Farm: Rural America Transformed*, Philadelphia: University of Pennsylvania Press, pp. 192–206.

Dewey, A. (1962) *Peasant Marketing in Java*, New York: Free Press of Glencoe.

Donque, G. (1965–66) Le zoma de Tananarive: etude géographique d'un marché urbain. *Madagascar: Revue de Géographie* 7: 93–147; 8: 150–273.

DuPuis, E.M. (2000) Not in my body: rBGH and the rise of organic milk. *Agriculture and Human Values* 17: 285–95.

Easterly, W. (2001) The lost decades: developing countries' stagnation in spite of policy reform 1980–1998. *Journal of Economic Growth* 6: 135–57.

Eaton, C. and Shepherd, A.W. (2001) *Contract Farming: Partnerships for Growth: A Guide*, Rome: Food and Agriculture Organization of the United Nations.

Echánove, F. (2001) Working under contract for the vegetable agroindustry in Mexico: a means of survival. *Culture and Agriculture* 23: 13–23.

Echánove, F. and Steffen, C. (2003) Coping with trade liberalization: the case of Mexican grain producers. *Culture and Agriculture* 25(2): 31–42.

Echánove, F. and Steffen, C. (2005) Agribusiness and farmers in Mexico: the importance of contractual relations. *Geographical Journal* 171: 166–76.

Ellis, F. (1988) *Peasant Economics: Farm Households and Agrarian Development*, Cambridge: Cambridge University Press.

Ellis, F. (1998) Household strategies and rural livelihood diversification. *Journal of Development Studies* 35: 1–38.

Ellis, F. (2005) Small-farms, livelihood diversification and rural–urban transitions: strategic issues for Sub-Saharan Africa, paper presented at the research workshop The Future of Small Farms, Wye, UK, 26–29 June.

Ellis, F. and Mdoe, N. (2003) Livelihoods and rural poverty reduction in Tanzania. *World Development* 31: 1367–84.

Ellis, F. and Freeman, A.H. (2004) Rural livelihoods and poverty reduction strategies in four African countries. *Journal of Development Studies* 40: 1–30.

Elvin, M. (1973) *The Pattern of the Chinese Past*, Stanford, CA: Stanford University Press.

Elvin, M. (2004) *The Retreat of the Elephants: An Environmental History of China*, New Haven, CT and London: Yale University Press.

Evans, N., Morris, C. and Winter, M. (2002) Conceptualizing agriculture: a critique of post-productivism as the new orthodoxy. *Progress in Human Geography* 26: 313–32.

Eyferth, J. (2003) How not to industrialize: observations from a village in Sichuan. *Journal of Peasant Studies* 30: 75–92.

Eyferth, J., Ho, P. and Vermeer, E.B. (2003) Introduction: the opening up of China's countryside. *Journal of Peasant Studies* 30: 1–17.

Fanfani, R. and Brasili, C. (2003) Regional differences in Chinese agriculture: results from the 1997 First National Agricultural Census. *Journal of Peasant Studies* 30: 18–44.

Faure, D. (1989) The lineage as a cultural invention: the case of the Pearl River delta. *Modern China* 15(1): 4–36.

Fei, Hsiao-Tung and Chang, Chih-I (1945) *Earthbound China: A Study of Rural Economy in Yunnan*, Chicago, IL: University of Chicago Press.

Frances, J. (2003) The role of gangmasters and gang labour in the UK food chain network, past and present. Memorandum submitted to the Select Committee on Environment, Food and Rural Affairs, UK Parliament 2003, *Minutes of Evidence*.

Francks, P. (1996) From peasant to entrepreneur in Italy and Japan. *Journal of Peasant Studies* 22(4): 699–709.

Francks, P. (2005) Multiple choices: rural household diversification and Japan's path to industrialization. *Journal of Agrarian Change* 5: 451–75.

Frank, A.G. (1967) *Capitalism and Underdevelopment in Latin America: Historical Studies of Chile and Brazil*, New York: Monthly Review Press.

Franklin, S.H. (1969) *The European Peasantry: The Final Phase*, London: Routledge.

Fraters, B., Hotsma, P.H., Langenberg, V.T., van Leeuwen, T.C., Mol, A.P., Oltshoorn, C.S., et al. (2004) *Agricultural Practice and Water Quality in the Netherlands in the 1992–2002 Period*, Bilthoven, the Netherlands: National Institute of Public Health and the Environment.

Friedmann, H. (1978a) Simple commodity production and wage labour on the American plains. *Journal of Peasant Studies* 6: 71–100.

Friedmann, H. (1978b) World market, state and family farm: social bases of household production in the era of wage labour. *Comparative Studies in Society and History* 20: 545–86.

Freidmann, H. (1980) Household production and the national economy: concepts for the analysis of agrarian formations. *Journal of Peasant Studies* 7: 158–84.

Friedmann, H. (1993) The political economy of food: a global crisis. *New Left Review* I 197: 29–57.

Fuller, A.M. (1990) From part-time farming to pluriactivity: a decade of change in rural Europe. *Journal of Rural Studies* 6: 362–73.

Garciá, A. (1970) Agrarian reform and social development in Bolivia, in R. Stavenhagen (ed.) *Agrarian Problems and Peasant Movements in Latin America*, New York: Doubleday Anchor, pp. 301–46.

García Fajardo, B., Nava Bernal, G. and González Esquivel, C. (2004) Local soil conservation technologies and their role in farmers' livelihoods in San Pablo Tlalchichilpa, State of Mexico. *PLEC News and Views* NS 5: 15–20.

Garnaut, J. and Lim-Applegate, H. (1998) *People in Farming*, ABARE Research Report 98.6, Canberra: Australian Bureau of Agricultural and Resource Economics.

Garnaut, J., Rasheed, C. and Rodriguez, G. (1999) *Farmers at Work: The Gender Division*, Canberra: Australian Bureau of Agricultural and Resource Economics.

Gasper, D. (1986) Distribution and development ethics: a tour, in R. Apthorpe and A. Krahl (eds) *Development Studies: Critique and Renewal*, Leiden: E.J. Brill, pp. 136–203.

Geertz, C. (1963) *Peddlers and Princes: Social Change and Economic Modernization in Two Indonesian Towns*, Chicago, IL and London: University of Chicago Press.

Geertz, C. (1965) *The Social History of an Indonesian Town*, New York and London: Free Press of Glencoe.

Gleissman, S.R. (1998) *Agroecology: Ecological Processes in Sustainable Agriculture*, Chelsea, MI: Ann Arbor Press.

Glover, D. and Kusterer, K. (1990) *Small Farmers, Big Business Contract Farming and Rural Development*, London: Palgrave Macmillan.

González Casanova, P. (1970) *Democracy in Mexico*, New York: Oxford University Press (originally published as *La Democracia en México*, Mexico, DF: Ediciones Era (1965)).

Good, C.M. (1970) *Rural Markets and Trade in East Africa*, Research Paper No. 128, Chicago, IL: University of Chicago.

Goodman, D. (2004) Rural Europe redux? Reflections on alternative agro-food networks and paradigm change. *Sociologia Ruralis* 44(1): 3–16.

Goodman, D. and Redclift, M. (1981) *From Peasant to Proletarian: Capitalist Development and Agrarian Transitions*, Oxford: Basil Blackwell.

Goodman, D. and Redclift, M. (1985) Capitalism, petty commodity production and the farm enterprise. *Sociologia Ruralis* 25: 231–47.

Goodman, D. and Watts, M. (1994) Reconfiguring the rural or fording the divide? Capitalist restructuring and the global agro-food system. *Journal of Peasant Studies* 22: 1–49.

Gorton, P., Lowe, P. and Zellei, A. (2005) Pre-accession Europeanisation: the strategic realignment of the environmental policy systems of Lithuania, Poland and Slovakia towards agricultural pollution in preparation for EU membership. *Sociologia Ruralis* 45: 202–23.

Gray, J. (1997) The Common Agricultural Policy and the re-invention of the rural in the European Community. *Sociologia Ruralis* 40: 30–52.

Gray, J. (1998) Family farms in the Scottish borders: a practical definition by hill sheep farmers. *Journal of Rural Studies* 14: 341–56.

Gray, J. (2006) Mao in perspective. *China Quarterly* 187: 659–79.

Griffin, K. (1974) *The Political Economy of Agrarian Change*, London: Macmillan.

Griffin, K., Khan, A.R. and Ickowitz, A. (2002) Poverty and the distribution of land. *Journal of Agrarian Change* 2: 279–330.

Griffin, K., Khan, A.R. and Ickowitz, A. (2004) In defence of neo-classical neo-populism. *Journal of Agrarian Change* 4: 361–86.

Grigg, D.B. (1984) The agricultural revolution in western Europe, in T.P. Bayliss-Smith and S. Wanmali (eds) *Understanding Green Revolutions: Agrarian Change and Development Planning in South Asia*, Cambridge: Cambridge University Press, pp. 1–17.

Guillet, D. (1997) Water-demand management and farmer-managed irrigation systems. *Culture and Agriculture* 19(1/2): 1–5.

Guillet, D. (1998) Rethinking legal pluralism: local law and state law in the evolution of water property rights in northwestern Spain. *Comparative Studies in Society and History* 40(1): 42–70.

Guillet, D. (2006) Rethinking irrigation efficiency: chain irrigation in northwestern Spain. *Human Ecology* 34(3): 305–29.

Gultiano, S., Urich, P., Balbarino, E. and Saz, E. (2004) *Population Dynamics, Land Availability and Adapting Land Tenure Systems: Philippines, A Case Study*, Rome and Paris: FAO and Committee for International Cooperation in National Research in Demography (CICRED).

Guo, X. (2001) It's all a matter of hats: rural urbanization in south-west China. *Journal of Peasant Studies* 29: 109–28.

Guthman, J. (2004) The trouble with 'organic lite' in California: a rejoinder to the 'conventionalization' debate. *Sociologia Ruralis* 44(3): 301–16.

Guyer, J.I. (1997) *An African Niche Economy: Farming to Feed Ibadan, 1968–1988*, Edinburgh: Edinburgh University Press.

Gwynne, R.N. (2003) Transnational capitalism and local transformation in Chile. *Tijdschrift voor Economische en Sociale Geografie* 94: 310–21.

Gwynne, R.N. (2006) Export orientation and enterprise development: a comparison of New Zealand and Chilean wine production. *Tijdschrift voor Economische en Sociale Geografie* 97: 138–56.

Gyasi, E.A. and Asante, W.J. (2005) Aspects of resource tenure that conserve biodiversity, in E.A. Gyasi, G. Kranjac-Berisavljevic, E.T. Blay and W. Oduro (eds) *Managing Agrodiversity in the Traditional Way: Lessons from West Africa in Sustainable Use of Biodiversity and Related Natural Resources*, Tokyo: United Nations University Press, pp. 217–27.

Gyasi, E.A., Odruro, W., Kranjac-Berisavljevic, G., Dittoh, J.S. and Asante, W. (2003) Ghana, in H. Brookfield, H. Parsons and M. Brookfield (eds) *Agrodiversity: Learning from Farmers across the World*, Tokyo: United Nations University Press, pp. 79–109.

Gyasi, E.A., Kranjac-Berisavljevic, G., Blay, E.T. and Oduro, W. (eds) (2005) *Managing Agrodiversity in the Traditional Way: Lessons from West Africa in Sustainable Use of Biodiversity and Related Natural Resources*, Tokyo: United Nations University Press,

Hacker, L.M. and Kendrick, B.B. (1949) *The United States since 1865*, fourth edition, New York: Appleton-Century-Crofts.

Haenn, N. (2006) The changing and enduring ejido: a state and regional examination of Mexico's land tenure counter reforms. *Land Use Policy* 23: 136–46.

Hall, A. (2003) Canadian agricultural policy: liberal, global and sustainable, in J. Adams (ed.) *Fighting for the Farm: Rural America Transformed*, Philadelphia: University of Pennsylvania Press, pp. 209–28.

Hall, A. and Mogyorody, V. (2001) Organic farmers in Ontario: an examination of the conventionalization argument. *Sociologia Ruralis* 41(4): 399–422.

Hardin, G. (1968) The tragedy of the commons. *Science* 162: 1243–8.

Hardjono, J. (1987) *Land, Labour and Livelihood in a West Java Village*, Yogyakarta: Gadjah Mada University Press.

Harnecker, M. (2002) *Landless People: Building a Social Movement*, translation of *Sin Tierra: Construyendo Movimento Social*, Madrid, Siglo XXI. Online. Available at <www.rebelion.org/harnecker/landless300802.pdf> (accessed 1 May 2006).

Haroon Akram-Lodhi, A. (2005) Vietnam's agriculture: processes of rich peasant accumulation and mechanisms of social differentiation. *Journal of Agrarian Change* 5(1): 73–116.

Harvey, N. (1998) *The Chiapas Rebellion: The Struggle for Land and Democracy*, Durham, NC: Duke University Press.

Hayek, F.A. von (1976) *The Mirage of Social Justice*, London: Routledge.

Hecht, S.B. (2005) Soybeans, development and conservation on the Amazon frontier. *Development and Change* 36(2): 375–404.

Helfand, S.M. (1999) The political economy of agricultural policy in Brazil: decision making and influence from 1964 to 1992. *Latin American Research Review* 34(2): 3–41.

Hill, B. (1993) The 'myth' of the family farm: defining the family farm and assessing its importance in the European Community. *Journal of Rural Studies* 9: 359–70.

Hill, P. (1963) *The Migrant Cocoa Farmers of Southern Ghana*, London: Cambridge University Press.

Hinton, W. (1966) *Fanshen*, New York: Monthly Review Press.

Hinton, W. (1983) *Shenfan*, New York: Random House.

Hinton, W. (1990) *The Great Reversal: The Privatization of China 1978–1989*, New York: Monthly Review Press.

Ho, S.P.S. (1995) Rural non-agricultural development in post-reform China: growth, development patterns and issues. *Pacific Affairs* 68: 360–91.

Hobsbawm, E.J. (1968) *Industry and Empire: An Economic History of Britain since 1750*, London: Weidenfeld and Nicolson.

Hobsbawm, E.J. (1994) *Age of Extremes: The Short Twentieth Century 1914–1991*, London: Michael Joseph.

Hodder, B.W. (1967) The markets of Ibadan, in P.C. Lloyd, A.L. Mabogunje and B. Awe (eds) *The City of Ibadan: A Symposium on its Structure and Development*, Cambridge: Cambridge University Press, pp. 173–90.

Hoggart, K. and Mendoza, C. (1999) African immigrant workers in Spanish agriculture. *Sociologia Ruralis* 39(4): 538–62.

Hoggart, K. and Paniagua, A. (2001a) What rural restructuring? *Journal of Rural Studies* 17: 41–62.

Hoggart, K. and Paniagua, A. (2001b) The restructuring of rural Spain. *Journal of Rural Studies* 17: 63–80.

Huang, P.C.C. (1985) *The Peasant Economy and Social Change in North China*, Stanford, CA: Stanford University Press.

Hughes, I. (1973) Stone-age trade in the New Guinea inland: historical geography without history, in H. Brookfield (ed.) *The Pacific in Transition: Geographical Perspectives on Adaptation and Change*, London: Edward Arnold, pp. 97–126.

Hunt, D. (1979) Chayanov's model of peasant household resource allocation. *Journal of Peasant Studies* 6(3): 247–85.

IFAD (International Fund for Agricultural Development) (2001) *Rural Poverty Report*, Oxford: Oxford University Press.

ILO [Singer, H. and Jolly, R. with others] (1972) *Employment, Incomes and Equality: A Strategy for Increasing Productive Employment in Kenya*, Geneva: International Labour Organization.

ILO (2002) Agriculture and employment, Geneva: International Labour Organization. Online. Available at <http://www.ilo.org/public/english/dialogue/sector/sectors/agri/emp.htm> (accessed 12 Dec. 2006).

IUCN (1980) *World Conservation Strategy: Living Resources Conservation for Sustainable Development*, Gland, Switzerland: International Union for the Conservation of Nature.

Izcara-Palacios, S.P. (1998) Farmers and the implementation of the EU Nitrates Directive in Spain. *Sociologia Ruralis* 38: 146–62.

Jacks, G.V. and Whyte, R.O. (1939) *Vanishing Lands: A World Survey of Soil Erosion*, New York: Doubleday Anchor.

Janssen, M.A. (2007) Coordination in irrigation systems: an analysis of the Lansing–Kremer model of Bali. *Agricultural Systems* 93(1): 170–90.

Jayne, T.S., Mather, D. and Mghenyi, E. (2005) Smallholder farming in difficult circumstances: policy issues for Africa, paper presented at research workshop on The Future of Small Farms, Wye, UK, 26–29 June.

Jervell, A.M. (1999) Changing patterns of family farming and pluriactivity. *Sociologia Ruralis* 39: 101–16.

Jodha, N.S. (1987) A case study of the degradation of common property resources in India, in P. Blaikie and H. Brookfield (eds) *Land Degradation and Society*, London: Routledge, pp. 196–207.

Johnsen, S. (2003) Contingency revealed: New Zealand farmers' experiences of agricultural restructuring. *Sociologia Ruralis* 43: 128–51.

Johnsen, S. (2004) The redefinition of family farming: agricultural restructuring and farm adjustment in Waihemo, New Zealand. *Journal of Rural Studies* 20: 419–32.

Johnson, E.A.J. (1970) *The Organization of Space in Developing Countries*, Cambridge, MA: Harvard University Press.

Johnson, R., Manning, J. and Scrimgeour, F. (2006) Public investment in agricultural research in New Zealand, 1990–2005. *Farm Policy Journal* 3(1): 41–51.

Juul, K. (2005) Transhumance, tubes and telephones: drought-related migration as a process of innovation, in Q. Gausset, M.A. Whyte and T. Birch-Thomsen (eds) *Beyond Territory and Scarcity: Exploring Conflicts over Natural Resource Management*, Stockholm: Nordiska Afrikainstitutet, pp. 112–34.

Kaihura, F. (2005) Viewing the Millennium Development Goals from the lake zone of northern Tanzania. *PLEC News and Views* NS 8: 5–10.

Kaihura, F. and Stocking, M. (eds) (2003) *Agricultural Biodiversity in Smallholder Farms of East Africa*, Tokyo: United Nations University Press.

Karaczun, Z., Lowe, P. and Zellei, A. (2003) *The Challenge of the Nitrate Directive to Acceding Countries: A Comparative Analysis of Poland, Lithuania and Slovakia*, vol. 2 of F. Gatzweiler and K. Hagedom (eds) *Institutional Change in Central and Eastern European Agriculture and Environment*, Berlin: Humboldt University for FAO.

Kautsky, K. (1899) *Die Agrafrage*, Stuttgart: Dietz; trans. Pete Burgess (1988) *The Agrarian Question*, Winchester, MA: Zwan.

Kay, C. (2005) Reflections on rural poverty in Latin America. *European Journal of Development Studies* 17: 317–46.

Kay, C. (2006) Raúl Prebisch, in D. Simon (ed.) *Fifty Key Thinkers on Development*, London: Routledge, pp. 199–205.

Kerkvliet, B.J.T. and Selden, M. (1998) Agrarian transformations in China and Vietnam. *China Journal* 40: 37–58.

Kerven, C. (1992) *Customary Commerce: A Historical Reassessment of Pastoral Livestock Marketing in Africa*, ODI Agricultural Occasional Paper 15, London: Overseas Development Institute.

Kinsella, J., Wilson, S., de Jong, F. and Renting, H. (2000) Pluriactivity as a livelihood strategy in Irish farm households and its role in rural development. *Sociologia Ruralis* 40: 481–96.

Kipnis, A. (1995) Within and against peasantness: backwardness and filiality in rural China. *Comparative Studies in Society and History* 37: 110–35.

Kirwan, J. (2004) Alternative strategies in the UK agro-food system interrogating the alterity of farmers' markets. *Sociologia Ruralis* 44(4): 395–415.

Kleijn, D. and Sutherland, W.J. (2003) How effective are European agri-environmental schemes in conserving and promoting biodiversity? *Journal of Applied Ecology* 40: 947–69.

Knickel, K. and Renting, H. (2000) Methodological and conceptual issues in the study of multifunctionality and rural development. *Sociologia Ruralis* 40(4): 512–28.

Kritzinger, A., Barrientos, S. and Rossouw, H. (2004) Global production and flexible employment in South African horticulture: experiences of contract workers in fruit exports. *Sociologia Ruralis* 44(1): 17–39.

Kueh, Y.Y. (2006) Mao and agriculture in China's industrialization: three antitheses in a 50-year perspective. *China Quarterly* 187: 700–23.

Ladurie, E. Le Roy (1976) *The Peasants of Languedoc*, trans. by J. Day, Urbana, IL: University of Illinois Press.

Lamartine-Yates, P. (1940) *Food Production in Western Europe: An Economic Survey of Agriculture in Six Countries*, London: Longmans, Green.

Lamartine-Yates, P. (1960) *Food, Land and Manpower in Western Europe*, London: Macmillan.

Lansing, J.S. (1987) Balinese water temples and the management of irrigation. *American Anthropologist* 89: 326–41.

Lansing, J.S. (1991) *Priests and Programmers: Technologies of Power in the Engineered Landscape of Bali*, Princeton, NJ: Princeton University Press.

Lansing, J.S. and Kremer, J.N. (1993) Emergent properties of Balinese water temple networks: coadaptation on a rugged fitness landscape. *American Anthropologist* 95(1): 97–114.

Lansing, J.S. and Miller, J.H. (2005) Cooperation, games and ecological feedback: some insights from Bali. *Current Anthropology* 46(2): 328–33.

Latacz-Lohmann, U. and Hodge, I. (2003) European agri-environmental policy for the twenty-first century. *Australian Journal of Agricultural and Resource Economics* 47: 123–39.

Lawrence, P. (1955) Land tenure among the Garia, Canberra: Australian National University, Social Science Monograph 4; reprinted in H.I. Hogbin and P. Lawrence (eds) (1967) *Studies in New Guinea Land Tenure*, Sydney: Sydney University Press.

Lawson, T. (2003) *Reorienting Economics*, London: Routledge.

Leach, E.R. (1961) *Pul Eliya: A Village in Ceylon, A Study of Land Tenure and Kinship*, Cambridge: Cambridge University Press.

Leaf, M. (1984) *Song of Hope: The Green Revolution in a Punjab Village*, New Brunswick, NJ: Rutgers University Press.

Lehmann, D. (1986) Two paths of agrarian capitalism, or a critique of Chayanovian Marxism. *Comparative Studies in Society and History* 28: 601–27.

Lehmann D. (1996) Peasants, in A. Kuper and J. Kuper (eds) *The Social Science Encyclopedia*, London and New York: Routledge, pp. 595–6.

Lenin, V.I. (1899) *The Development of Capitalism in Russia: The Process of the Formation of a Home Market for Large-scale Industry*, Moscow: Foreign Languages Publishing House; translated and reprinted in V.I. Lenin (1961) *Collected Works*, fourth English edition, Moscow: Foreign Languages Publishing House.

Leopold, A. (1966) *A Sand Country Almanac with Other Essays on Conservation from Round River*, New York: Oxford University Press.

Levi-Strauss, C. (1966) *The Savage Mind*, London: Weidenfeld and Nicholson.

Lewis, W.A. (1954) Economic development with unlimited supplies of labour. *Manchester School of Economics and Social Science* 22: 139–91.

Li, H. and Rozelle, S. (2003) Privatizing rural China: insider privatization, innovative contracts and the performance of township enterprises. *China Quarterly* 176: 981–1005.

Li, L.C. (2006) Differentiated actors: central–local politics in China's rural tax reforms. *Modern Asian Studies* 40(1): 151–74.

Liendo Vera, I. (1997) Estructura de la Tenancia y su distribución especial en el Estado de México. In G.R. Herrejón, A.A. Hernández, L. Gonzáles Diaz and C. Arriaga Jordán (eds) *Investigación Para el Desarollo Rural: Diez Años de Experiencia del CICA*, Toluca, México: Centro de Investigación en Ciencias Agropecuarias, Universidad Autónoma del Estado de México, pp. 386–423.

Lipton, M. (1968) The theory of the 'optimizing peasant'. *Journal of Development Studies* 4: 327–51.

Lipton, M. (1977) *Why Poor People Stay Poor: A Study of Urban Bias in World Development*, London: Temple Smith.

Liu Ching (1964) *The Builders*, Peking [Beijing]: Foreign Languages Press.

Liu, L. and Murphy, R. (2006) Lineage networks, land conflict and rural migration in late-socialist China. *Journal of Rural Studies* 33(4): 612–45.

Mabogunje, A.L. (1967) The morphology of Ibadan, in P.C. Lloyd, A.L. Mabogunje and B. Awe (eds) *The City of Ibadan: A Symposium on its Structure and Development*, Cambridge: Cambridge University Press, pp. 35–56.

Mabogunje, A.L. (1968) *Urbanization in Nigeria*, London: University of London Press.

Mabry, J.B. (1996) *Canals and Communities: Small-scale Irrigation Systems*, Tucson: University of Arizona Press.

McCarthy, J. (2005) Rural geography: multifunctional rural geographies – reactionary or radical? *Progress in Human Geography* 29(6): 773–82.

MacDonald, D., Crabtree, J.R., Wiesinger, G., Dax, T., Stamou, N., Fleury, P., *et al.* (2000) Agricultural abandonment in mountain areas of Europe: environmental consequences and policy response. *Journal of Environmental Management* 59: 47–69.

McGee, T.G. and Greenberg, C. (1992) The emergence of extended metropolitan regions in ASEAN. *ASEAN Economic Bulletin* 9: 22–44.

McGregor, C.J. and Warren, C.R. (2006) Adopting sustainable farm management practices within a nitrate vulnerable zone in Scotland: the view from the farm. *Agriculture, Ecosystems and Environment* 113: 108–19.

McKay, B. and Acheson, J. (1987) *The Question of the Commons: The Culture and Ecology of Communal Resources*, Tucson: University of Arizona Press.

McMichael, P. (1994) *The Global Restructuring of Agro-Food Systems*, Ithaca, NY: Cornell University Press.

Malaysia (1976) *Third Malaysia Plan 1976–81*, Kuala Lumpur: National Printing Department.

Malaysia (1991) *Sixth Malaysia Plan 1991–95*, Kuala Lumpur: National Printing Department.

Malinowski, B. (1922) *Argonauts of the Western Pacific: An Account of Native Enterprise and Adventure in the Archipelago of Melanesian New Guinea*, London: Routledge and Kegan Paul.

Mann, S. and Dickinson, J. (1978) Obstacles to the development of a capitalist agriculture. *Journal of Peasant Studies* 5(4): 466–81.

Mariategui, J.C. (1928) *Siete Ensayos de Interpretacion de la Realidad Peruana*, trans. by M. Urquidi (1971) as *Seven Interpretative Essays on Peruvian Reality*, Austin, Texas: University of Texas Press.

Marsden, T. (1999) Rural futures: the consumption countryside and its regulation. *Sociologia Ruralis* 39: 501–20.

Marsden, T., Whatmore, S., Munton, R. and Little, J. (1986) The restructuring process and economic centrality in capitalist agriculture. *Journal of Rural Studies* 2: 271–80.

Marsden, T., Banks, J. and Bristow, G. (2000) Food supply chain approaches: exploring their role in rural development. *Sociologia Ruralis* 40(4): 424–38.

Martins, J. de Souza. (2002) Representing the peasantry? Struggles for/about land in Brazil. *Journal of Peasant Studies* 29(3/4): 300–35.

Marx, K. (1867) *Das Kapital: Kritik der Politichen Oekonomie*, vol. 1, Hamburg: Verlag von Otto Meissner.

Marx, K. (1965) *Pre-capitalist Economic Formations*, trans. J. Cohen, E.J. Hobsbawm (ed. and introduction by), New York: International Publishers.

Mather, A.S., Hill, G. and Nijnik, M. (2006) Post-productivism and rural land use: cul de sac or challenge for theorization? *Journal of Rural Studies* 22: 441–55.

Mayrand, K., Dionne, S., Paquin, M., Ortega, G.A. and Guadarrama Marrón (2003) *Reengineering of Agricultural Policy in OECD Countries; Trends and Policy Implications for Mexico*, Montreal and Mexico City: Unisféra International Centre and Centro Mexicano de Derecho Ambiental.

Meinzen-Dick, R., Brown, L., Feldstein, H. and Quisumbing, A. (1997) Gender, property rights and natural resources. *World Development* 25: 1303–15.

Melichar, E. (1977) Some current aspects of agricultural finance and banking in the United States. *American Journal of Agricultural Economics* 59: 967–72.

Middleton, J. (1798) *View of the Agriculture of Middlesex*, London: Board of Agriculture.

Morris, C. and Winter, M. (1999) Integrated farming systems: the third way for European agriculture? *Land Use Policy* 16: 193–205.

Murdoch, J., Marsden, T. and Banks, J. (2000) Quality, nature and embeddedness: some theoretical considerations in the context of the food sector. *Economic Geography* 76(2): 107–25.

Murray, W.E. (2002) From dependency to reform and back again: the Chilean peasantry during the twentieth century. *Journal of Peasant Studies* 6: 190–227.

Myers, N. (1988) Threatened biotas: 'hotspots' in tropical forests. *Environmentalist* 8(3): 1–20.

Myers, N., Mittermeier, R., Mittermeier, C., de Fonseca, G. and Kent, J. (2000) Biodiversity hotspots for conservation priorities. *Nature* 403: 842–3.

Nadal, A. (2000) *The Environmental and Social Impacts of Economic Liberalization of Corn Production in Mexico*, Oxford and Gland (Switzerland): Oxfam and WWF.

Nadel, S.F. (1942) *A Black Byzantium: The Kingdom of Nupe in Nigeria*, Oxford: Oxford University Press.

Nagatsuka, T. (1989) *The Soil: A Portrait of Rural Life in Meiji Japan*, trans. and with an introduction by A. Waswo, London and New York: Routledge.

Navarro, Z. (2005) Transforming rights into social practices? The landless movement and land reform in Brazil. *IDS Bulletin* 36(1): 129–37.

Netting, R. McC. (1968) *Hill Farmers of Nigeria: Cultural Ecology of the Kofyar of the Jos Plateau*, Seattle: University of Washington Press.

Netting, R. McC. (1993) *Smallholders, Householders: Farm Families and the Ecology of Intensive, Sustainable Agriculture*, Stanford, CA: Stanford University Press.

Nuijten, M. (2003) Family property and the limits of intervention: the Article 27 reforms and the PROCEDE programme in Mexico. *Development and Change* 34: 475–97.

O'Keefe, M. (2005) Fresh food retailing: a growth story. *Farm Policy Journal* 2(1): 28–36.

O'Rourke, E. (2006) Changes in agriculture and the environment in an upland region of the Massif Central, France. *Environmental Science and Policy* 9: 370–5.

Oi, J. (1989) *State and Peasant in Contemporary China*, Berkeley and Los Angeles: University of California Press.

Ong, L. (2006) The political economy of township government debt, township enterprises and rural financial institutions in China. *China Quarterly* 188: 377–400.

Orden, D. (2003) U.S. agricultural policy: the 2002 farm bill and the WTO Doha Round proposal, Trade and Macroeconomics Division Discussion Paper 109, Washington: International Food Policy Research Institute.

Östberg, W. (1995) *Land is Coming Up: The Burunge of Central Tanzania and Their Environment*, Stockholm: Department of Social Anthropology, Stockholm University.

Ostrom, E. (1990) *Governing the Commons*, Cambridge: Cambridge University Press.

Padel, S. (2001) Conversion to organic farming: a typical example of the diffusion of an innovation? *Sociologia Ruralis* 41(1): 40–61.

Pallot, J. and Nefedova, T. (2003) Trajectories in people's farming in Moscow oblast during the post-socialist transformation. *Journal of Rural Studies* 19: 345–62.

Pasquini, M.W. and Alexander, M.J. (2005) Soil fertility management strategies on the Jos plateau: the need for integrating 'empirical' and 'scientific' knowledge in agricultural development. *Geographical Journal* 171(2): 112–24.

Pavelis, G.A. (1983) Conservation capital in the United States, 1935–1980. *Journal of Soil and Water Conservation* 38: 455–8.

Petras, J. (2006) 'Centre-left' regimes in Latin America: history repeating itself as farce? *Journal of Peasant Studies* 33: 278–303.

Pimm, S.L., Ayres, M., Balmford, A., Branch, G., Brandon, K., Brooks, T., *et al.* (2001) Can we defy nature's end? *Science* 293(5538): 2207–8.

Pinedo-Vasquez, M., McGrath, D.G. and Ximenes, T. (2003) Brazil (Amazonia), in H. Brookfield, H. Parsons and M. Brookfield (eds) *Agrodiversity: Learning from Farmers across the World*, Tokyo: United Nations University Press, pp. 41–78.

Platteau, J-P. (1994) Behind the market stage where real societies exist – Part I: The role of public and private order institutions. *Journal of Development Studies* 30: 533–77.

Polanyi, K. (1944) *The Great Transformation*, Boston, MA: Beacon Press.

Polanyi, K., Arensberg, C.M. and Pearson, H.W. (eds) (1957) *Trade and Market in the Early Empires*, New York: Free Press of Glencoe.

Policy Commission on the Future of Farming and Food [England] (2002) *Farming and Food: A Sustainable Future* (Chair: Sir Donald Curry), London: Cabinet Office. Online. Available at <http://www.cabinetoffice.gov.uk/farming> (accessed 13 Nov. 2006).

Porter, D. and Craig, D. (2004) The third way and the third world: poverty reduction and social inclusion in the rise of 'inclusive' liberalism. *Review of International Political Economy* 12: 387–423.

Porter, G. and Phillips-Howard, K. (1997) Comparing contracts: an evaluation of contract farming schemes in Africa. *World Development* 25: 227–38.

Prazan, J., Ratinger, T., Krumalova, V., Lowe, P. and Zellei, A. (2003) *Maintaining High Nature Value Landscapes in an Enlarged Europe: A Comparative Analysis of the Czech Republic, Hungary and Slovenia*, vol. 1 of F. Gatzweiler and K. Hagedom (eds) *Institutional Change in Central and Eastern European Agriculture and Environment*, Berlin: Humboldt University for FAO.

Prebisch, R. (1950) *El Desarollo Ecónomico de la América Latina y Algunos de sus Principales Problemas*, Santiago: Comisión Ecónomica para América Latina, trans. as *The Economic Development of Latin America and its Principal Problems*, Lake Success, NY: United Nations.

Preston, D.A. (1989) Too busy to farm: under-utilisation of farm land in central Java. *Journal of Development Studies* 26: 43–57.

Pritchard, B. (2006) The political construction of free trade visions: the geo-politics and geo-economics of Australian beef exporting. *Agriculture and Human Values* 23: 37–50.

Productivity Commission (2004) *Impacts of Native Vegetation and Biodiversity Regulations*, Report No. 29, Melbourne: Australian Productivity Commission.

Prugl, E. (2004) Gender orders in German agriculture: from the patriarchal welfare state to liberal environmentalism. *Sociologia Ruralis* 44(4): 349–72.

Rackham, O. (1986) *The History of the Countryside*, London: J.M. Dent.

Radu, M. (2004) Mexico's Zapatistas: another failed revolution. Online. Available at <http://www.frontpagemag.com/> (accessed 4 Sept. 2006).

Reardon, T., Timmer, P., Barrett, C.R. and Berdegué, J. (2003) The rise of supermarkets in Africa, Asia and Latin America. *American Journal of Agricultural Economics* 85: 1140–6.

Reichert, T. (2006) *A Closer Look at EU Agricultural Subsidies: Developing Modification Criteria*, Hamm and Berlin: Arbeitsgemeinschaft Bäuerliche Landwirtchaft and Germanwatch.

Reid, J.D. (1977) The theory of share tenancy revisited – again. *Journal of Political Economy* 85: 403–7.

Reij, C., Scoones, I. and Toulmin, C. (1996) *Sustaining the Soil: Indigenous Soil and Water Conservation in Africa*, London: Earthscan.

Renting, H., Marsden, T.K. and Banks, J. (2003) Understanding alternative food networks: exploring the role of short food supply chains in rural development. *Environment and Planning A* 35: 393–411.

Restrepo Fernández, I. (1976) La transferencia de recursos a la agricultura. *El Economista Mexicano* 3: 21–38.

Ricardo Flores Mago'n [autonomous municipality in rebellion] (2002) Ricardo Flores Mago'n is not going to allow this expulsion. Online. Available at <http://flag.blackened. net/revolt/mexico/ezln> (accessed 9 Sept. 2006).

Rigby, D. and Cáceres, D. (2000) Organic farming and the sustainability of agricultural systems. *Agricultural Systems* 68: 21–40.

Rigg, J. (2001) *More than the Soil: Rural Change in Southeast Asia*, Harlow, UK: Pearson Education.

Rigg, J. (2003) *Southeast Asia: The Human Landscape of Modernization and Development*, second edition, London and New York: Routledge.

Rigg, J. and Ritchie, M. (2002) Production, consumption and imagination in rural Thailand. *Journal of Rural Studies* 18: 359–71.

Roberts, R. (1996) Recasting the 'agrarian question': the reproduction of family farming on the southern high plains. *Economic Geography* 72: 398–415.

Robertson, A.F. (1980) On sharecropping. *Man* 15: 411–29.

Rochelau, D. and Edmunds, D. (1997) Women, men and trees: gender, power and property in forest and agrarian landscapes. *World Development* 25: 1351–71.

Rodrik, D. (2006) Goodbye Washington consensus, Hello Washington confusion? A review of the World Bank's 'Economic Growth in the 1990s: Learning from a Decade of Reform'. *Journal of Economic Literature, American Economic Association* 44(4): 973–87.

Rogaly, B. (2006) Intensification of work-place regimes in British agriculture: the role of migrant workers, Sussex Migration Working Paper 36, Falmer: University of Sussex.

Rogaly, B. and Rafique, A. (2003) Struggling to save cash: seasonal migration and vulnerability in West Bengal, India. *Development and Change* 34(4): 659–81.

Rohde, R.F., Moleele, N.M., Mphale, M., Allsop, N., Chanda, R., Hoffmen, M.T., *et al.* (2006) Dynamics of grazing policy and practice: environmental and social impacts in three communal areas of southern Africa. *Environmental Science and Policy* 9: 302–16.

Rosset, P.M. and Altieri, M. (1997) Agroecology versus input substitution: a fundamental contradiction of sustainable agriculture. *Society and Natural Resources* 10(3): 283–95.

Rostow, W.W. (1960) *The Stages of Economic Growth: A Non-Communist Manifesto*, Cambridge: Cambridge University Press.

Rowland, W. (1973) *The Plot to Save the World: The Life and Times of the Stockholm Conference on the Human Environment*, Toronto and Vancouver: Clarke Irwin.

Sahlins, M. (1972) *Stone-Age Economics*, Chicago, IL: Aldine.

Salais, R. and Storper, M. (1992) The four worlds of contemporary industry. *Cambridge Journal of Economics* 16: 169–93.

Sanders, R. (2000) *Prospects for Sustainable Development in the Chinese Countryside: The Political Economy of Chinese Ecological Agriculture*, Aldershot, UK: Ashgate Publishing.

Sanders, R. (2006) A market road to sustainable agriculture? Ecological agriculture, green food and organic agriculture in China. *Development and Change* 37: 201–26.

Sarfo-Mensah, P. and Oduro, W. (2003) The dynamics of population change and land management in the savanna transition zone of Ghana. *PLEC News and Views* NS 3: 3–10.

Saugeres, L. (2002) Of tractors and men: masculinity, technology and power in a French farming community. *Sociologia Ruralis* 42(2): 143–59.

Schultz, T. (1964) *Transforming Traditional Agriculture*, New Haven, CT: Yale University Press.

Schmitt, G. (1991) Why is the agriculture of advanced Western economies still organized by family farms? Will this continue to be so in the future? *European Review of Agricultural Economics* 18: 443–458.

Scoones, I. (1998) *Sustainable Rural Livelihoods: A Framework for Analysis*, IDS Working Paper No. 72, Brighton, UK: Institute of Development Studies.

Scoones, I. and Wolmer, W. (2006) *Livestock, Disease, Trade and Markets: Policy Choices for the Livestock Sector in Africa*, Working Paper 269, Brighton, UK: Institute of Development Studies.

Scott, J.C. (1976) *The Moral Economy of the Peasant: Rebellion and Subsistence in Southeast Asia*, New Haven, CT: Yale University Press.

Scott, J.C. (1985) *Weapons of the Weak: Everyday Forms of Peasant Resistance*, New Haven, CT and London: Yale University Press.

Seers, D. (1979) The congruence of Marxism and other neoclassical doctrines, in K.Q. Hill (ed.) *Toward a New Strategy of Development: A Rothko Chapel Colloquium*, New York: Pergamon, pp. 1–17.

Sender, J. and Johnson, D. (2004) Searching for a weapon of mass production in rural Africa: unconvincing arguments for land reform. *Journal of Agrarian Change* 4: 142–64.

Serova, E. (2002) Evolution of the farm structure in Russia's agriculture: background and perspectives. Online. Available at <http://www.iet.ru/afe/conferences/prague.pdf> (accessed 4 Sept. 2006).

Shortall, S. (2002) Gendered agricultural and rural restructuring: a case study of Northern Ireland. *Sociologia Ruralis* 42(2): 160–75.

Shreck, A. (2005) Resistance, redistribution and power in the Fair Trade banana initiative. *Agriculture and Human Values* 22: 17–29.

Shucksmith, M. and Winter, M. (1990) The politics of pluriactivity in Britain. *Journal of Rural Studies* 6: 429–35.

Shue, V. (1980) *Peasant China in Transition: The Dynamics of Development Toward Socialism, 1949–1956*, Berkeley: University of California Press.

Silcock, T.H. and Fisk, E.K. (eds) (1963) *The Political Economy of Independent Malaya: A Case Study in Development*, Canberra: Australian National University Press.

Simpson, R. (2004) The evolution of French agri-environmental schemes from territorial land management contracts (CTEs) to contracts for sustainable agriculture (CADs). The Countryside Agency. Online. Available at <http://www.countryside.gov.uk/Images/French%20agri-env%20schemes%20from%20CTEs%20to%20CADsFinal_tcm2-26663.pdf> (accessed 6 March 2006).

Singer, H.W. (1950) The distribution of gains between investing and borrowing countries. *American Economic Review* 40: 473–85.

Singh, S. (2002) Contracting out solutions: political economy of contract farming in the Indian Punjab. *World Development* 30: 1621–38.

Siu, H. (1989) *Agents and Victims in South China: Accomplices in Rural Revolution*, New Haven, CT: Yale University Press.

Siu, H. and Faure, D. (1995) Introduction, in H. Siu and D. Faure (eds) *Down to Earth: The Territorial Bond in South China*, Palo Alto, CA: Stanford University Press, pp. 1–18.

Sivakumar, S.S. (2001) The unfinished Narodnik agenda: Chayanov, Marxism and marginalism revisited. *Journal of Peasant Studies* 29: 31–60.

Skinner, G.W. (1964) Marketing and social structure in rural China, Part I. *Journal of Asian Studies* 24(1): 3–44.

Smil, V. (1984) *The Bad Earth: Environmental Degradation in China*, Armonk, NY: M.S. Sharpe.

Smil, V. (2001) *Feeding the World: A Challenge for the Twenty-first Century*, Cambridge, MA: MIT Press.

Smith, A. (1776) *An Inquiry into the Nature and Causes of the Wealth of Nations*, Edinburgh: W. Strahan & T. Cadell.

Smith, E. and Marsden, T. (2004) Exploring the 'limits to growth' in UK organics: beyond the statistical image. *Journal of Rural Studies* 20: 345–57.

Smith, T.C. (1959) *The Agrarian Origins of Modern Japan*, Stanford, CA: Stanford University Press.

Smith, T.C. (1988) *Native Sources of Japanese Industrialization, 1750–1920*, Berkeley and Los Angeles: University of California Press.

Smith, W. (1973) *Innovation and Diffusion – A Supply Oriented Example: Hybrid Grain Corn in Quebec*, Department of Geography Discussion Paper 8, Columbus, Ohio: Ohio State University.

Smith, W. (1994) If you haven't got any socks, you can't pull them up: a preliminary report on the impact of agricultural policy reforms on land management strategies in the Dannevirke region [New Zealand], unpublished report for Lindsay Saunders, Landcare.

Smith, W. and Montgomery, H. (2003) Revolution or evolution? New Zealand agriculture since 1984. *Geojournal* 59: 107–18.

Smithers, J. and Furman, M. (2003) Environmental farm planning in Ontario: exploring participation and the endurance of change. *Land Use Policy* 20: 343–56.

Smithers, J., Joseph, A.E. and Armstrong, M. (2005) Across the divide (?): reconciling farm and town views of agriculture–community linkages. *Journal of Rural Studies* 21: 281–95.

Sobels, J., Curtis, A. and Lockie, S. (2001) The role of Landcare group networks in rural Australia: exploring the contribution of social capital. *Journal of Rural Studies* 17: 265–76.

Solidarité Paysanne: *Bulletin Mensuel de la Confèdération Paysanne Nord-Pas de Calais* (Janvier 2007), No. 306, Spécial élections.

South Africa (1955) Summary of the Report of the Commission for the Socio-economic Development of the Bantu Areas within the Union of South Africa, Pretoria: Government Printer.

Stamp, L.D. and Beaver, S.H. (1941) *The British Isles: A Geographic and Economic Survey*, third edition, London: Longmans, Green.

Stavenhagen, R. (1970) *Agrarian Problems and Peasant Movements in Latin America*, New York: Anchor Books.

Stavenhagen, R. (2001) Mexico's unfinished symphony: the Zapatista movement. Online. Available at <http://w.w.w.incore.ulst.ac.uk/services/ecrd/Stavenhagen.pdf> (accessed 4 Sept. 2006).

Stockdale, A., Findlay, A. and Short, D. (2000) The repopulation of rural Scotland: opportunity and threat. *Journal of Rural Studies* 16: 243–57.

Stocking, M. (1978) Remarkable erosion in central Rhodesia. *Proceedings of the Geographical Association of Rhodesia* 11: 42–56.

Stocking, M. (1996) Soil erosion: breaking new ground, in M. Leach and R. Mearns (eds) *The Lie of the Land: Challenging Received Wisdom on the African Environment*, London: International African Institute, pp. 140–54.

Stocking, M. and Murnaghan, N. (2003) *Handbook for the Field Assessment of Land Degradation*, London: Earthscan.

Stone, G.D. (1997) *Settlement Ecology: The Social Organization of Kofyar Agriculture*, Tucson: University of Arizona Press.

Tarrant, J. (1980) *Food Policies*, Chichester: John Wiley.

Tichelar, M. (2004) The Scott Report and the Labour Party: the protection of the countryside during the Second World War. *Rural History* 15(2): 167–87.

Tiffen, M. (2003) Transition in sub-Saharan Africa: agriculture, urbanization and income growth. *World Development* 37: 1343–66.

Tiffen, M. (2006) Urbanization: impacts on the evolution of 'mixed farming' systems in sub-Saharan Africa. *Experimental Agriculture* 42: 259–87.

Tong, S., Hall, C.A.S. and Wang, H. (2003) Land use change in rice, wheat and maize production in China (1961–1998). *Agriculture, Ecosystems and Environment* 95: 523–36.

Toulmin, C. (1992) *Cattle, Women and Wells: Managing Household Survival in the Sahel*, Oxford: Clarendon Press.

Toulmin, C. and Guèyé, B. (2003) Transformations in West African Agriculture and the role of family farms, Drylands Programme Issue Paper, no. 123, London: IIED.

Toulmin, C. and Guèyé, B. (2005) Is there a future for family farming in West Africa? *IDS Bulletin* 36(2): 23–29.

Twyman, C., Sporton, D. and Thomas, D.S.G. (2004) 'Where is the life in farming?' The viability of smallholder farming on the margins of the Kalahari, southern Africa. *Geoforum* 35(1): 69–85.

UK Parliament (2003) Select Committee on Environment, Food and Rural Affairs *Fourteenth Report*.

UK Treasury and Defra (2005) *A Vision for the Common Agricultural Policy*, London: H.M. Treasury and Department for Environment, Food and Rural Affairs.

UNEP (United Nations Environment Programme) (2002) *Africa Environment Outlook: Past, Present and Future Perspectives*, Stevenage, Herts, UK: Earthprint for the United Nations Environment Programme.

Unruh, J.D. (2006) Land tenure and the 'evidence landscape' in developing countries. *Annals of the Association of American Geographers* 96(4): 754–72.

Vakulabharanam, V. (2005) Growth and distress in a South Indian peasant economy during the era of economic liberalization. *Journal of Development Studies* 41(6): 971–7.

Van der Haar, G. (2005) Land reform, the state, and the Zapatista uprising in Chiapas. *Journal of Peasant Studies* 32: 484–507.

Van der Ploeg, J.D. (2000) Revitalizing agriculture: farming economically as starting ground for rural development. *Sociologia Ruralis* 40(4): 497–511.

Van der Ploeg, J.D. and Renting, H. (2000) Impact and potential: a comparative review of European development practices. *Sociologia Ruralis* 40(4): 529–43.

Van der Ploeg, J.D. and Renting, H. (2004) Behind the 'redux': a rejoinder to David Goodman. *Sociologia Ruralis* 44(2): 233–42.

Van der Ploeg, J.D., Renting, H., Brunori, G., Knickel, K., Mannion, J., Marsden, T., *et al.* (2000) Rural development: from practices and policies toward theory. *Sociologia Ruralis* 40(4): 391–408.

Villafuerte Solis, D. (2005) Rural Chiapas ten years after the armed uprising of 1994: an economic overview. *Journal of Peasant Studies* 32: 461–83.

Villers, S. (1998) Italy: single currency for a plural country. *Conjoncture* 21: 21–7.

Voeks, R.A. (1998) Ethnobotanical knowledge and environmental risk: foragers and farmers in northern Borneo, in K.S. Zimmerer and K.R. Young (eds) *Nature's Geography: New Lessons for Conservation in Developing Countries*, Madison: University of Wisconsin Press, pp. 307–26.

Vogeler, I. (1981) *The Myth of the Family Farm: Agribusiness Dominance of U.S. Agriculture*, Boulder, CO: Westview Press.

Wade, R. (1979) Fast growth and slow development in southern Italy, in D. Seers, B. Schaffer and M-L. Kiljunen (eds) *Underdeveloped Europe: Studies in Core–Periphery Relations*, Hassocks, Sussex, UK: Harvester Press, pp. 197–221.

Waldhardt, R., Simmering, D. and Albrecht, H. (2003) Floristic diversity at the habitat scale in the agricultural landscapes of central Europe. *Agriculture, Ecosystems and Environment* 98: 79–85.

Walford, N. (2003) A past and a future for diversification on farms? Some evidence from large-scale, commercial farms in South East England. *Geografiska Annaler, Series B: Human Geography* 85(1): 51–62.

Walker, K.L.M. (2006) 'Gangster capitalism' and peasant protest in China: the last twenty years. *Journal of Peasant Studies* 33: 1–33.

Wanmali, S. (1981) *Periodic Markets and Rural Development in India*, Delhi: B.R. Publishing.

Ward, B.E. (1960) Cash or credit crops? An examination of some implications of peasant commercial production to the multiplicity of traders and middlemen. *Economic Development and Cultural Change* 8: 148–63.

Warren, B. (1973) Imperialism and capitalist industrialization. *New Left Review* I 81: 1–33.

Warren, B. (1979) The postwar economic experience of the Third World, in K.Q. Hill (ed.) *Toward a New Strategy for Development*, New York: Pergamon, pp. 144–68.

Watson, A. and Findlay, C. (1999) *Food Security and Economic Reform: The Challenges Facing China's Grain Marketing System*, London and New York: Macmillan.

Watts, M. (1994) Life under contract: contract farming, agrarian restructuring, and flexible accumulation, in P.D. Little and M.J. Watts (eds) *Living under Contract: Contract*

Farming and Agrarian Transformation in Sub-Saharan Africa, Madison: University of Wisconsin Press, pp. 21–77.

Wegren, S.K., O'Brien, D.J. and Patsiorkovski, V.V. (2003) Winners and losers in Russian agrarian reform. *Journal of Peasant Studies* 30: 1–29.

Welch, C. (2004) Brazilian cordiality and peasant mobilization, before, during and after military rule, paper presented at a symposium on the Cultures of Dictatorship: Historical Reflections on the Brazilian Golpe of 1964, University of Maryland, 14–16 October.

Welch, C. (2006) Movement histories: a preliminary historiography of Brazil's Landless Laborer Movement (MST). *Latin American Research Review* 41: 198–210.

Wilkie, R. (1971) *San Miguel: A Mexican Collective Ejido*, Stanford, CA: Stanford University Press.

Willis, S. and Campbell, H. (2004) The chestnut economy: the praxis of neo-peasantry in rural France. *Sociologia Ruralis* 44(3): 317–31.

Wilson, E.O. (ed.) (1988) *BioDiversity*, Washington, DC: National Academy Press.

Wilson, G.A. (2001) From productivism to post-productivism . . . and back again? Exploring the (un)changed natural and mental landscapes of European agriculture. *Transactions of the Institute of British Geographers* NS 26: 77–102.

Wilson, G.A. (2004) The Australian *Landcare* movement: towards 'post-productivst' rural governance? *Journal of Rural Studies* 20: 461–84.

Wilson, G.A. and Rigg, J. (2003) 'Post-productivist' agricultural regimes and the South: discordant concepts? *Progress in Human Geography* 27: 681–707.

Winarto, Y. (1995) State intervention and farmer creativity: integrated pest management among rice farmers in Subang, West Java. *Agriculture and Human Values* 12(3): 47–57.

Winarto, Y. (2004) *Seeds of Knowledge: The Beginning of Integrated Pest Management in Java*, New Haven, CT: Yale University Southeast Asia Studies.

Wiser, C.V. (1978) *Four Families of Karimpur*, Syracuse, NY: Syracuse University.

Wiskerke, J.S.C. (2003) On promising niches and constraining sociotechnical regimes: the case of Dutch wheat and bread. *Environment and Planning A* 35: 429–48.

Wolf, D.L. (1992) *Factory Daughters: Gender, Household Dynamics, and Rural Industrialization in Java*, Berkeley and Los Angeles: University of California Press.

Wolf, E.R. (1982) *Europe and the People without History*, Berkeley and Los Angeles: University of California Press.

Wood, D. and Lenné, J.M. (1999) *Agrobiodiversity: Characterization, Utilization and Management*, New York: CABI and Oxford University Press.

Wood, J.A. and Skaggs, R. (n.d.) *An Analysis of Farm Labor Contracting in New Mexico*, New Mexico Chile Task Force Report 18, Las Cruces, NM: University of New Mexico.

Woodgate, G. (1997) The Mexican municipality: tensions between indigenous knowledge, sustainable livelihoods and bureaucratic strategies, in G.R. Herrejón, A.A. Hernández, L. Gonzáles Diaz and C. Arriaga Jordán (eds) *Investigación Para el Desarollo Rural: Diez Años de Experiencia del CICA*, Toluca, México: Centro de Investigación en Ciencias Agropecuarias, Universidad Autónoma del Estado de México, pp. 226–58.

World Commission on Environment and Development (1987) *Our Common Future*, Oxford: Oxford University Press.

Wrigley, E.A. (1985) Urban growth and agricultural change: England and the continent in the early modern period. *Journal of Interdisciplinary History* 15(4): 683–728.

Wrigley, N. (2001) The consolidation wave in U.S. food retailing: a European perspective. *Agribusiness* 17: 489–513.

Wu, B. and Pretty, J. (2004) Social connectedness in marginal rural China: the case of farmer innovation circles in Zhidan, north Shaanxi. *Agriculture and Human Values* 21: 81–92.

Xiande, L. (2003) Rethinking the peasant burden: evidence from a Chinese village. *Journal of Peasant Studies* 30: 45–74.

Yang, C.K. (1959a) *The Chinese Family in the Communist Revolution*, Cambridge, MA: MIT Press.

Yang, C.K. (1959b) *A Chinese Village in Early-Communist Transition*, Cambridge, MA: MIT Press.

Yep, R. (2004) Can 'tax-for-fee' reform reduce rural tension in China? The process, progress and limitations. *China Quarterly* 177: 42–70.

Yin, R., Xu, J. and Li, Z. (2003) Building institutions for markets: experiences and lessons from China's rural forestry sector. *Environment, Development and Sustainability* 5: 333–51.

Yin, Shaoting (2001) *People and Forests: Yunnan Swidden Agriculture in Human–Ecological Perspective*, trans. by Magnus Fiskesjo, Kunming: Yunnan Education Publishing House.

Yoshihara, K. (1988) *The Rise of Ersatz Capitalism in Southeast Asia*, Singapore: Oxford University Press.

Young, H. and Burke, M. (2000) Competition and custom in economic contracts, Working Papers in Economics 428, Baltimore, MD: Johns Hopkins University Press.

Yunez Naude, A. (2003) The dismantling of CONASUPO, a Mexican state trader in agriculture. *World Economy* 26(1): 97–122.

Zhao, S.X-B. and Wong, K.K-K. (2002) The sustainability dilemma of China's township and village enterprises: an analysis from spatial and functional perspectives. *Journal of Rural Studies* 18: 257–73.

Zhou, K. (1996) *How the Farmers Changed China: Power of the People*, Boulder, CO: Westview Press.

Zhou, Y.Z., Sumner, D.A. and Lee, H. (2002) Part-time farming trends in China, in Ligang Song (ed.) *Dilemmas of China's Growth in the Twenty-first Century*, Canberra: Asia-Pacific Press, pp. 194–214.

Zhu, Y.Y., Wang, Y.Y., Chen, H.R. and Lu, B.R. (2003) Conserving traditional rice varieties through management for crop diversity. *Bioscience* 53(2): 158–62.

Zimmerer, K.S. (1996) *Changing Fortunes: Biodiversity and Peasant Livelihood in the Peruvian Andes*, Berkeley and Los Angeles: University of California Press.

Zimmerer, K.S. (2006) Geographical perspectives on globalization and environmental issues: the inner-connections of conservation, agriculture and livelihoods, in K. Zimmerer (ed.) *Globalization and the New Geographies of Conservation*, Chicago, IL: University of Chicago Press, pp. 1–43.

Index